Multiple Choice Questions in Anatomy

John Pegington FRCS

S A Courtauld Professor of Anatomy
The Department of Anatomy & Developmental Biology
University College London

Edward Arnold
A division of Hodder & Stoughton
LONDON MELBOURNE AUCKLAND

Preface

The time set aside in curricula for medical students to learn the language, facts and concepts of Topographical Anatomy has been drastically reduced during the past two decades. Course organisers have been forced to concentrate on basic facts and those parts of Anatomy specifically needed to prepare the student for the clinical subjects. Multiple – choice questions offer the student a rapid way of testing knowledge and a means of gauging the emphasis which should be placed on various topics. A collection of such questions also offers practice for undergraduate examinations, many of which are now found in this format.

While the reduction in learning time for Topographical Anatomy was taking place, the development of imaging techniques, computer tomography (CT), magnetic resonance imaging (MRI) and ultrasound (US) began to influence the type of Topographical Anatomy required in medical training. Nowadays, much more diagnosis is 'anatomically' based than a decade ago, and with the advance of imaging techniques this trend is likely to continue. Questions have therefore been introduced in this text for the student to gauge his skill at interpreting both radiographic and CT images.

When testing yourself on a set of questions in this book, give +1 for each correct answer, −1 for each incorrect response, and 0 for items you leave unanswered. The score for each question can therefore vary from +5 to −5 marks. The pass mark in multiple – choice question papers often puzzles candidates. It is, like other examinations, an arbitrary cut-off level. Multiple – choice question papers of 25–40 questions similar to those in this book have been in use at University College London for a number of years. The mean score at the final Anatomy examination is usually 45%–48% with a standard deviation of about 8–12. We set the pass mark at the mean score minus a standard deviation. This is not held as a rigid pass mark, however, for other factors such as the difficulty of the particular exam and the distribution curve of class marks are included in the formulation of the final pass mark. This usually means scaling the mark list upwards. Marks from the practical examination and a short-answer paper are also taken into account. For the purposes of this book, you can consider that the 'pass: fail' boundary lies in the 40%–45% range at your first attempt: if you can achieve scores in this range you have passed. On a second attempt, however, you should do considerably better.

It is extremely difficult to set a group of questions which do not have one or two ambiguities. At University College London the questions are examined by external examiners, and we also ask students themselves to make a note of any ambiguities on the MCQ paper. I would be grateful to hear from any readers with comments, criticisms, and errors.

I am deeply indebted to Dr David Edwards, Head of the Department of Radiology at University College Hospital for the radiographs and CT scans, Mr Mike Gilbert of the Photography Unit of the Department of Anatomy & Developmental Biology at UCL for the photographic work, and to Churchill Livingstone, Publishers, for allowing me to use the photographic illustrations of these radiographs and CT scans from 'Clinical Anatomy in Action'. Many thanks go to Miss Sue Pryer for spending so much time proof reading the text. Finally thanks to the staff of Edward Arnold for their help and encouragement during the production of this book.

John Pegington
London 1989

Contents

Preface 2

Thorax and Vertebral column 4

Abdomen 20

Pelvis and perineum 36

Head and neck 50

Upper limb 70

Lower limb 84

Thorax and Vertebral Column

1 The arch of the aorta is related to the
  (a) ligamentum arteriosum.
  (b) thoracic duct.
  (c) left superior intercostal vein.
  (d) vena azygos.
  (e) left recurrent laryngeal nerve.

2 The left lung
  (a) has a transverse fissure.
  (b) may have an azygos lobe.
  (c) has a section called the lingula.
  (d) is related, through pleura, to the left phrenic nerve.
  (e) reaches the level of the 8th rib in the mid-axillary line during quiet respiration.

3 The thoracic duct
  (a) enters the thorax behind the medial arcuate ligament.
  (b) is related to the oesophagus in the thorax.
  (c) drains lymph from the right lung.
  (d) receives lymph from the pelvis.
  (e) may receive lymph from the left side of the head and neck.

4 The internal thoracic artery
  (a) is a branch of the subclavian artery.
  (b) supplies blood to the female breast.
  (c) gives anterior intercostal branches.
  (d) gives branches to the thymus.
  (e) gives branches to the thyroid gland.

1 (a) **True** The ligamentum arteriosum extends between the concavity of the aortic arch and the left pulmonary artery. It is a remnant of the ductus arteriosus, a channel which conducted deoxygenated blood from the pulmonary artery to the aorta during fetal life.

(b) **True** The thoracic duct lies on the left side of the oesophagus at this level and both are crossed by the arch.

(c) **True** The vein runs on the left surface of the arch superficial to the left vagus and deep to the left phrenic nerve.

(d) **False** The vena azygos arches over the right lung root and has the oesophagus on its medial side.

(e) **True** The nerve hooks under the aortic arch: an alteration in the voice may be the first sign of a tumour in the region.

2 (a) **False** The transverse fissure is found only in the right lung.

(b) **False** During development of the right lung a small segment may become 'cut off' as it grows upwards past the azygos vein. This is known as an azygos lobe, and the fissure is usually visible on a chest radiograph.

(c) **True** A section of the upper lobe consisting of two bronchopulmonary segments is known as the lingula.

(d) **True** The nerve runs in front of the lung root.

(e) **True** The surface markings for the lower border of the lungs during quiet breathing are the 6th costal cartilage, 8th rib in the mid-axillary line, and the 10th thoracic spine.

3 (a) **False** The thoracic duct passes through the aortic opening in the diaphragm.

(b) **True** The duct ascends behind the oesophagus in the posterior mediastinum, and then inclines to the left in the superior mediastinum.

(c) **False** Lymph from the right lung drains to the right broncho-mediastinal trunk.

(d) **True** The duct drains lymph from all over the body, except from the right side of the head and neck and thoracic wall, the right lung, the right side of the heart and the right upper limb.

(e) **True** The left jugular lymph trunk often enters directly into the thoracic duct.

4 (a) **True** Other branches include the vertebral artery, thyrocervical and costocervical trunks, and the dorsal scapular artery.

(b) **True** Perforating branches of this vessel running through the 2nd, 3rd and 4th intercostal spaces to the female breast are particularly large.

(c) **True** These supply the upper six intercostal spacse (2 anterior intercostal vessels to each space).

(d) **True** The gland may also receive branches from the inferior thyroid artery.

(e) **False** The thyroid arteries arise from the external carotid and subclavian arteries. An additional branch sometimes arises from the aortic arch.

5 The tenth intercostal nerves
  (a) supply skin immediately above the symphysis pubis.
  (b) have lateral cutaneous branches.
  (c) contain parasympathetic fibres.
  (d) supply intercostal muscles.
  (e) supply abdominal wall musculature.

6 The articulation between
  (a) the first rib and the manubrium is synovial.
  (b) the second costal cartilage and the sternum is synovial.
  (c) the manubrium and sternum is a primary cartilaginous joint.
  (d) the tubercle of a typical rib and an adjacent transverse process is synovial.
  (e) a rib and its costal cartilage is a primary cartilaginous joint.

7 The anterior interventricular artery
  (a) supplies the SA node in over 60% of hearts.
  (b) arises from the anterior aortic sinus.
  (c) supplies blood to the interventricular septum.
  (d) travels in the coronary sulcus.
  (e) gives off left anterior ventricular arteries.

8 Concerning the conducting system of the heart,
  (a) the sinu-atrial node (SA node) is located in the atrial wall adjacent to the opening for the coronary sinus.
  (b) the atrioventricular bundle lies adjacent to the membranous part of the interventricular septum.
  (c) fibres of the right crus travel through the septomarginal trabecula (moderator band).
  (d) the atrioventricular node is supplied with blood by a branch of the right coronary artery in over 80% of hearts.
  (e) cardiac impulses originate in the SA node.

9 Features found when examining the interior of the right atrium include
  (a) a part of the wall derived from the septum primum.
  (b) a part of the wall derived from the free edge of the septum secundum.
  (c) the opening of the coronary sinus.
  (d) a valve of the inferior vena cava.
  (e) trabeculae carneae.

5 (a) **False** The 10th intercostal nerves supply skin at the level of the umbilicus.
(b) **True** All the intercostal nerves have both lateral and anterior cutaneous branches, although those of the 1st may be small.
(c) **False** Intercostal nerves have sympathetic fibres which innervate sweat glands, arrectores pilorum muscles and blood vessels.
(d) **True** Muscular branches supply the intercostal muscles of the 10th space. The collateral branch of the nerve is sensory in function (Gray), although it is sometimes said to be a motor branch (Last).
(e) **True** The rectus, obliques and transversus receive a supply from the 10th nerves.

6 (a) **False** This is a primary cartilaginous joint.
(b) **True** This joint lies at the level of the sternal angle.
(c) **False** The joint is a secondary cartilaginous type.
(d) **True** The joint between the head of a typical rib and two vertebral bodies is also synovial.
(e) **True** Another example of a primary cartilaginous joint is the junction between an epiphysis and the diaphysis of a bone.

7 (a) **False** The SA node is supplied by a branch of the right coronary artery in 65% (Hutchinson, 1978).
(b) **False** The artery is usually a branch of the left coronary artery.
(c) **True** Both anterior and posterior interventricular arteries supply the septum.
(d) **False** It travels in the interventricular groove.
(e) **True** These travel obliquely to the left ventricular wall. One may be large and arise directly from the LCA, and is then called the left diagonal artery.

8 (a) **False** It is found in the right atrial wall close to the root of the superior vena cava.
(b) **True** Care must be taken during surgery in this region not to injure the bundle.
(c) **True** The septomarginal trabecula is often found crossing the cavity of the right ventricle from the septal to the opposite wall.
(d) **False** The node is supplied by a branch of the right coronary artery in 80% (Hutchinson, 1987).
(e) **True** The rate is usually about 72 per minute.

9 (a) **True** The floor of the fossa ovalis is derived from septum primum.
(b) **True** The limbus fossa ovalis is derived from the free edge of the septum secundum.
(c) **True** The coronary sinus opens into the atrial wall just below the fossa ovalis.
(d) **True** This valve is a developmental remnant and is functionless in the adult.
(e) **False** These are found on the ventricular walls.

10 Use the posteroanterior film of the chest to decide which of the following statements are true and which are false.

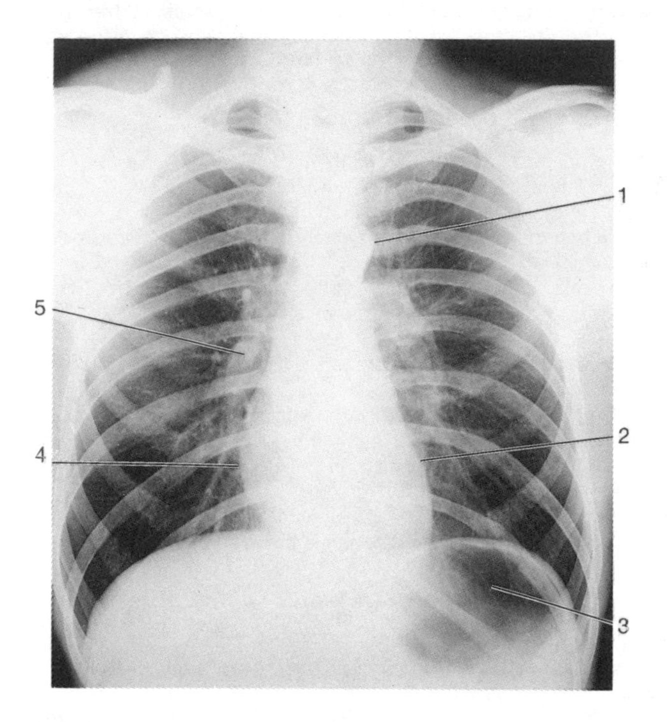

Fig. 1.1 Posteroanterior chest film. By kind permission of Churchill Livingstone Publishers. From *Clinical Anatomy in Action*, Volume 3: Pegington.

(a) The radiopaque bulge '1' is made by the aorta.
(b) The heart profile at '2' is produced by the left atrium.
(c) The radiolucent area '3' is produced by air in the lower lobe of the left lung.
(d) The heart profile at '4' is produced by the right atrium.
(e) The opacity marked '5' is made by the right principal bronchus.

11 The adult oesophagus
(a) is 20–25 cm long.
(b) begins at the level of the 6th cervical vertebra.
(c) is constricted by the left principal bronchus.
(d) is related, through pericardium, to the right atrium.
(e) crosses in front of the descending aorta.

10 (a) **True** It is known by radiologists as the aortic 'knuckle'.

(b) **False** The profile is made by the edge of the left ventricle.

(c) **False** This radiolucent area lies below the diaphragm, and is produced by air in the stomach; it is the 'gastric air bubble'.

(d) **True** Above this, the right profile is produced by the superior vena cava.

(e) **False** The air passages do not produce the hilar shadows. These are vascular markings made by pulmonary arteries and veins.

11 (a) **True** When examining the oesophagus with an oesophagoscope the cardiac orifice is usually about 40 cm from the patient's incisor teeth.

(b) **True** This is the level at which the oesophagus may be assumed to begin when examining a radiograph of the neck.

(c) **True** Normal constrictions may sometimes also be found at its commencement, the level of the aortic arch, and the passage of the oesophagus through the diaphragm.

(d) **False** The oesophagus is related to the left atrium through pericardium. An enlarged left atrium indents the outline of the oesophagus on a barium swallow. Mitral stenosis is one such cause of a large left atrium.

(e) **True** Traced from above, the oesophagus inclines to the left, crossing in front of the aorta.

12 Use the CT of the upper thorax in Fig. 1.2 to decide which of the following statements are true and which are false. The arteries contain contrast.

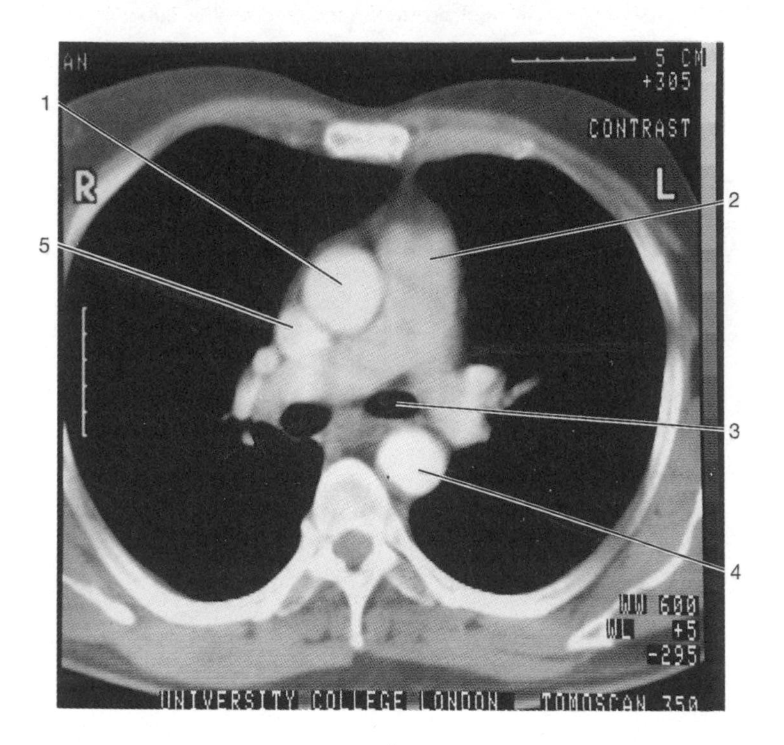

Fig. 1.2 CT of chest. By kind permission of Churchill Livingstone, Publishers. From *Clinical Anatomy in Action*, Volume 3: Pegington.

(a) The opaque circular shadow '1' is the azygos arch.
(b) The opacity labelled '2' is the pulmonary trunk.
(c) The black shadow labelled '3' is the trachea.
(d) The opaque circular shadow '4' is the descending thoracic aorta.
(e) The circular shadow labelled '5' is the superior vena cava.

13 The right sympathetic trunk
(a) lies in front of the neck of the first right rib.
(b) contributes fibres to the deep cardiac plexus.
(c) contains fibres which are responsible for dilatation of the right pupil.
(d) is continuous with the lumbar section of the trunk behind the lateral arcuate ligament.
(e) contributes preganglionic fibres to ganglia in the abdomen.

12 (a) **False** This is the ascending aorta.
(b) **True** The section is taken at the level of bifurcation of the trunk.
(c) **False** There are two black holes and these are the left and right principal bronchi.
(d) **True** The vessel lies just to the left of the vertebral body (probably T.5).
(e) **True** The student should remember that '5' is on the right-hand side of the subject – we are looking UPWARDS at a transverse section of the patient.

13 (a) **True** It is often the cervicothoracic (stellate) ganglion which lies in this position.
(b) **True** Cardiac branches are said to arise from all three cervical sympathetic ganglia and pass to the deep cardiac plexus. On the left cardiac branches also arise, but those from the superior cervical ganglion are said to run to the superficial cardiac plexus.
(c) **True** Fibres in the white ramus of the stellate ganglion ascend in the trunk to the head. If they are cut, Horner's syndrome is produced on the damaged side – a constricted pupil, ptosis, absence of sweating on the face, and enophthalmos.
(d) **False** The subcostal neurovascular bundle runs behind this ligament: the sympathetic trunk lies behind the medial arcuate ligament.
(e) **True** The splanchnic nerves consist of preganglionic fibres which pass to ganglia in the abdomen.

14 The aortic opening of the diaphragm
(a) lies at the level of L.2.
(b) is bounded by the left and right crura.
(c) transmits the vagus nerves.
(d) transmits the vena azygos.
(e) is situated in the central tendon.

15 Structures concerned in the normal development of the diaphragm include
(a) the septum transversum.
(b) the fourth cervical myotomes.
(c) the pleuropericardial membrane.
(d) mesoderm of the dorsal body wall.
(e) the mesentery of the oesophagus.

16 Concerning the state of development of the aortic arches just before birth.
(a) the ductus arteriosus is derived from the sixth left aortic arch.
(b) the first aortic arch artery has disappeared.
(c) the right recurrent laryngeal nerve hooks around a derivative of the sixth aortic arch.
(d) the fourth left arch has formed part of the adult aorta.
(e) the fifth arch arteries have disappeared.

14 (a) **False** The aortic opening lies at the level of T.12, the oesophageal opening at T.10 and the caval opening at T.8.
(b) **True** The crura are united in front of the aorta by the median arcuate ligament.
(c) **False** The vagus nerves leave the thorax through the oesophageal opening in the diaphragm.
(d) **True** The vena azygos usually enters the thorax through the aortic opening.
(e) **False** It is the caval opening that is situated in the tendinous part of the diaphragm.

15 (a) **True** The sternal and costal parts are derived from the septum transversum; a gap between these two parts is known as a foramen of Morgagni.
(b) **True** These invade the septum transversum and account for the nerve supply of the diaphragm (phrenic nerve C3, 4 and 5).
(c) **False** It is the pleuroperitoneal membranes that take a small part in the formation of the diaphragm.
(d) **True** An abnormal foramen (of Bochdalek) is sometimes found between the central tendon and the lumbar section of the diaphragm. It is usually left-sided.
(e) **True** This forms the diaphragm between the oesophageal and aortic openings.

16 (a) **True** The ventral portion of the sixth left arch is absorbed into the pulmonary trunk, whilst the dorsal portion becomes the ductus.
(b) **True** The first disappears entirely and the dorsal end of the second arch artery becomes the stapedial artery.
(c) **False** The dorsal portion of the sixth right arch disappears and the recurrent laryngeal nerve then hooks around the fourth arch artery (subclavian).
(d) **True** The fourth left arch forms much of the definitive aortic arch.
(e) **True** This arch disappears on both sides.

17 The atlas has
(a) a long spinous process.
(b) foramina transversaria.
(c) the vertebral artery running above its posterior arch.
(d) the first cervical spinal nerve running below its posterior arch.
(e) a secondary cartilaginous joint between its anterior arch and the dens (odontoid peg).

18 During a lumbar puncture the needle passes
(a) through an interspinous ligament.
(b) through dura.
(c) between the spinous processes of L1 and L2.
(d) through pia.
(e) into the lumbar cistern.

19 The intervertebral disc between the 5th lumbar vertebra and the sacrum in the adult
(a) contains a nucleus pulposus.
(b) is related posteriorly to the spinal cord.
(c) is related in front to the abdominal aorta.
(d) is a secondary cartilaginous joint.
(e) gives attachment to the posterior longitudinal ligament.

20 The veins of the internal vertebral venous plexus (of Batson)
(a) are valveless.
(b) communicate with venous plexuses in the cranial cavity.
(c) lie in the subdural space.
(d) communicate with intercostal veins.
(e) accept blood from basivertebral veins.

17 (a) **False** The atlas does not have a spinous process, only a posterior tubercle.
(b) **True** The vertebral arteries run through these foramina.
(c) **True** The vertebral artery runs from the foramen transversarium, over the posterior arch of the atlas, and into the vertebral canal.
(d) **False** Like the vertebral artery, the first cervical spinal nerve also runs on the upper surface of the posterior arch.
(e) **False** The joint is synovial: it may be affected by rheumatoid arthritis.

18 (a) **True** The needle is passed between adjacent spines, through the ligament.
(b) **True** Once through dura and arachnoid the needle point enters CSF.
(c) **False** This is too high and injury to the conus medullaris may occur. The correct level is between the spinous processes of L3 and L4 (or L4 and L5).
(d) **False** At the level of the puncture the pia is only associated with the filum terminale.
(e) **True** The pool of CSF in the lower lumbosacral region is called the lumbar cistern. It contains the roots of the lower spinal nerves (cauda equina) and the filum terminale.

19 (a) **True** All intervertebral discs contain a nucleus; that of L5-S1 is subject to compression by the weight of the upper body. It is often involved in the condition of 'slipped' disc, the nucleus being partly extruded through a damaged annulus fibrosus.
(b) **False** The spinal cord usually ends at the level of L1–L2 in the adult.
(c) **False** The aorta has bifurcated at the level of L4.
(d) **True** An intervertebral disc is a secondary cartilaginous type of joint.
(e) **True** The fibres of the posterior longitudinal ligament are firmly adherent to the annulus fibrosus.

20 (a) **True** They have few if any valves, and blood may therefore pass into the plexus from segmental veins or vice versa.
(b) **True** The basilar and occipital plexuses communicate with the vertebral plexus, thus establishing a venous outflow from structures in the cranial cavity apart from the IJV.
(c) **False** The plexus lies in the extradural tissues.
(d) **True** The segmental spinal veins in the thoracic region make this communication. Cells from a cancer of the breast may enter intercostal veins, spinal veins, the vertebral plexus, and then enter a vertebral body where they form a secondary deposit.
(e) **True** Basivertebral veins drain blood from the vertebral bodies.

21 Use the picture of a radiograph of the lumbar vertebral column in fig. 1.3 to decide which of the following statements are true and which are false.

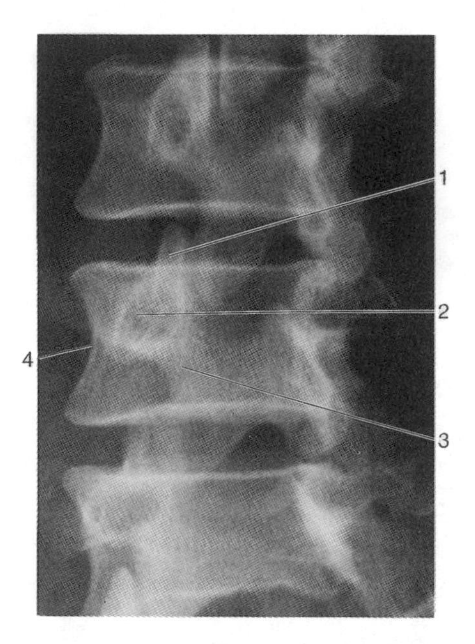

Fig. 1.3 Radiograph of lumbar vertebral column. By kind permission of Churchill Livingstone, Publishers. From *Clinical Anatomy in Action*, Volume 1: Pegington.

(a) This is a lateral view of the lumbar column.
(b) The projection marked '1' is a superior articular facet.
(c) The ring-like opacity marked '2' is the pedicle.
(d) The narrow opaque section of the vertebra marked '3' is the inferior articular facet.
(e) The shape of the vertebral body '4' indicates that it belongs to the fifth lumbar vertebra.

22 Features of the fourth and fifth cervical vertebrae include,
(a) foramina for the vertebral arteries.
(b) uncovertebral joints between adjacent bodies.
(c) mamillary processes.
(d) foramina for vertebral veins.
(e) facet joints lying in the sagittal plane.

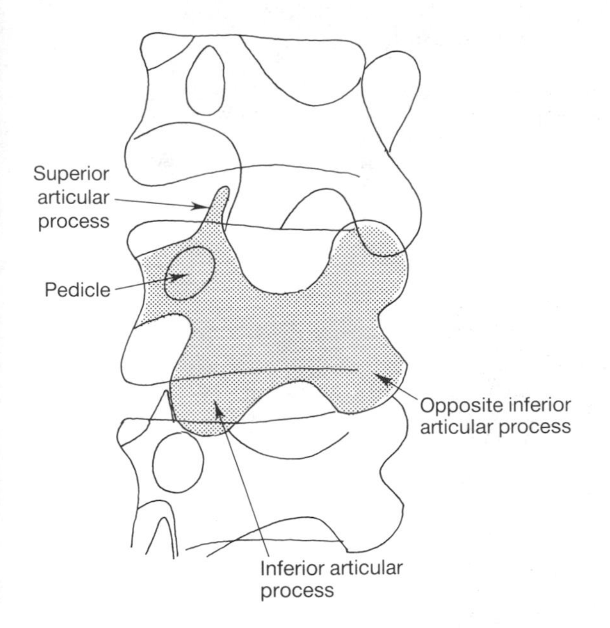

Fig. 1.4 Tracing from the radiograph of the lumbar spine.

21 (a) **False** It looks as though there is a 'Scottie' dog on the film, (Fig. 1.4). This film is taken with an oblique projection.

(b) **True** The 'ear' of the dog is produced by the superior articular facets.

(c) **False** The 'eye' of the dog is the transverse process seen end on.

(d) **False** This is the pars interarticularis and it is here that a defect sometimes occurs, especially in L5. This is known as spondylolysis. When this occurs there is a radiloucent 'collar' around the dog's neck.

(e) **False** The vertical length of the anterior surface of the 5th lumbar vertebra is greater than its posterior length. Thus it is shaped like a wedge. The other lumbar bodies have equal anterior and posterior vertical heights.

22 (a) **True** The foramina transversaria transmit the vertebral arteries. The vertebral arteries do not usually run through the foramina of C7, but this is variable.

(b) **True** These small joints are found at the sides of the typical cervical vertebral bodies, and although synovial at first, tend to fuse during adult life.

(c) **False** Mamillary processes are features of lumbar vertebrae.

(d) **True** Vertebral veins begin, not within the cranial cavity, but in the suboccipital region. They travel downwards through successive foramina transversaria to those of the sixth vertebra, leave and enter the brachiocephalic veins.

(e) **False** The flat facets of typical cervical vertebrae are best seen on a lateral radiograph, and slope at about 45 degrees to the coronal plane. The superior articular process faces upwards and its partner, the inferior process of the vertebra above, faces downwards.

23 Disc protrusions in the neck which compress the 5th and 6th cervical spinal nerves is likely to result in
(a) paralysis of the small muscles of the hand.
(b) altered skin sensation along the ulnar border of the forearm.
(c) loss of sweating on the mid-palm.
(d) loss of the biceps reflex.
(e) paralysis of the deltoid.

24 Use the diagram to decide which of the following statements are true and which are false. The dotted line represents a sympathetic preganglionic fibre, the line of crosses is a postganglionic sympathetic fibre.

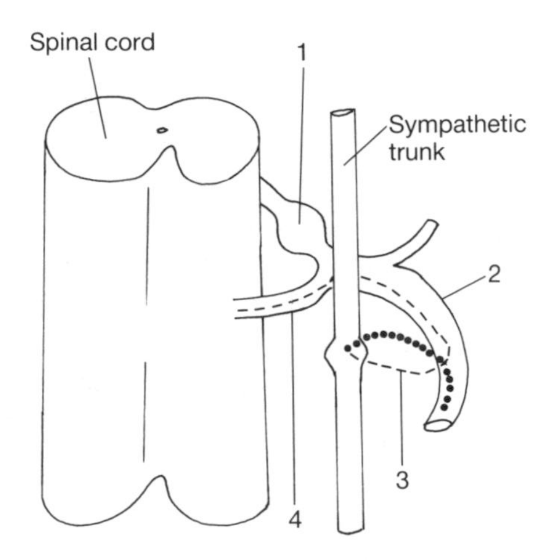

Fig. 1.5 A section of the spinal cord and left sympathetic trunk.

(a) The ganglion labelled '1' contains autonomic synapses.
(b) The nerve '2' is the ventral root of a spinal nerve.
(c) The small connection '3' is a grey ramus.
(d) The section labelled '4' contains sensory fibres.
(e) The diagram is likely to represent a cervical segment of the spinal cord and a cervical spinal nerve.

25 Concerning the spinal meninges
(a) the extradural space lies between the dura and the periosteum of the vertebral column.
(b) the ligamenta denticulata are projections of pia mater.
(c) the spinal dura is continuous with the cranial dura.
(d) a dural sheath extends along each spinal nerve.
(e) the spinal dura ends at the level of the L2 in the adult.

23 (a) **False** T1 is responsible for the supply of the intrinsic muscles of the hand.
(b) **False** Dermatomes C5 and C6 are found on the outer aspect of the arm.
(c) **False** Loss of sweating would indicate damage to the sympathetic supply. It is C7 which is responsible for the supply of the skin of the palm, and damage to this spinal segment would result in loss of sensation and sweating.
(d) **True** Spinal nerves C5 and C6 supply the biceps.
(e) **True** C5 and C6 supply fibres to the deltoid through the axillary nerve.

24 (a) **False** The spinal ganglion contains cell bodies of sensory fibres.
(b) **False** This is the ventral RAMUS, not a root.
(c) **False** This is the white ramus, containing myelinated pre-ganglionic fibres. It is usually found distal to the grey ramus on the ventral ramus.
(d) **False** The ventral ROOT contains motor fibres – in this case these include sympathetic motor fibres.
(e) **False** Sympathetic preganglionic fibres only leave the cord between T1 and L2 or L3, so this cannot be a picture of a cervical segment.

25 (a) **True** Anaesthetic may be introduced into this space, usually called the 'epidural' space by clinicians.
(b) **True** These are found between ventral and dorsal roots and extend from the pia covering the spinal cord to the dura mater. They support the spinal cord.
(c) **True** The spinal and cranial dura are continuous through the foramen magnum. The 'outer' layer of cranial dura is the endosteum of the cranial bones.
(d) **True** The dural sheaths of the spinal nerves fuse with the nerves just beyond the intervertebral foramina.
(e) **False** Although the spinal cord ends here, the dural tube reaches S2.

Abdomen

26 The spermatic cord
(a) is surrounded by tunica vaginalis.
(b) contains the iliohypogastric nerve.
(c) contains fibres of the transversus abdominis and internal oblique muscles.
(d) contains a branch of the inferior epigastric artery.
(e) contains the peritoneal sac of a direct inguinal hernia.

27 Posterior relations of the pancreas include
(a) the lesser sac.
(b) the portal vein.
(c) the splenic vein.
(d) the duodenum.
(e) the aorta.

28 Posterior relations of the second part of the duodenum include
(a) the transverse mesocolon.
(b) the hilum of the right kidney.
(c) the portal vein.
(d) the gastroduodenal artery.
(e) the right renal vein.

29 The right suprarenal gland
(a) receives blood from a branch of the right renal artery.
(b) is related to the inferior vena cava.
(c) drains its blood into the inferior vena cava.
(d) is related to the diaphragm.
(e) lies behind the lesser sac.

26 (a) **False** There is a double layer of tunica around the testis, and this may be connected to the peritoneum by a remnant of the processus vaginalis.
(b) **False** The only nerves contained within the spermatic cord are sympathetic nerves and the genital branch of the genitofemoral nerve.
(c) **True** These muscular fibres form the cremaster muscle.
(d) **True** The cremasteric artery is usually a branch of the inferior epigastric artery.
(e) **False** A direct hernia pushes through the posterior wall of the inguinal canal, medial to the deep ring. An indirect inguinal hernia, on the other hand, passes along a patent processus vaginalis within the cord.

27 (a) **False** The lesser sac lies in front of the pancreas.
(b) **True** The formation of the portal vein by the union of the superior mesenteric and splenic veins takes place behind the neck of the pancreas.
(c) **True** The splenic vein travels behind the body of the pancreas.
(d) **False** The pancreas lies in the concavity of the duodenum.
(e) **True** The aorta lies behind the body of the pancreas.

28 (a) **False** The transverse mesocolon lies in front of the second part of the duodenum.
(b) **True** The hilum of the right kidney, with the renal vein, pelvis and renal artery lie behind the second part of the duodenum.
(c) **False** The portal vein lies behind the pancreas and first part of the duodenum.
(d) **False** The gastroduodenal artery is an important posterior relation of the FIRST part of the duodenum, and is the artery which may be eroded by a posterior duodenal ulcer.
(e) **True** This is usually the most anterior of the structures in the hilum of the kidney.

29 (a) **True** Each suprarenal gland receives vessels from the renal artery, the aorta and the inferior phrenic artery.
(b) **True** The right gland is partly tucked behind the inferior vena cava, at the back of the liver.
(c) **True** The left suprarenal vein usually drains into the left renal vein. The pattern for both veins is not constant.
(d) **True** The right gland is related to both the bare area of the liver and the diaphragm.
(e) **False** It is the left suprarenal gland that lies behind the lesser sac.

30 The stomach
(a) receives vessels from the splenic artery.
(b) receives vessels from the superior mesenteric artery.
(c) receives fibres from both vagus nerves.
(d) drains blood to the portal system of veins.
(e) lies in front of the lesser sac (omental bursa).

31 The gall-bladder
(a) fundus lies at the level of the ninth right costal cartilage.
(b) usually receives blood from a branch of the right hepatic artery.
(c) lies partly between the peritoneal leaves of the lesser omentum.
(d) concentrates the bile about 100 times.
(e) is outlined during an intravenous cholangiogram.

32 The following are branches of the superior mesenteric artery.
(a) The ileocolic artery.
(b) The left colic artery.
(c) The superior rectal artery.
(d) The inferior pancreaticoduodenal artery.
(e) The gastroduodenal artery.

33 The portal vein
(a) contains many valves.
(b) is formed by the union of the splenic and inferior mesenteric veins.
(c) is closely related to the uncinate process of the pancreas.
(d) is separated from the inferior vena cava by peritoneum in part of its course.
(e) drains blood from the kidneys.

34 The vermiform appendix
(a) arises from the anterior surface of the caecum.
(b) is supplied with blood by a branch of the inferior mesenteric artery.
(c) contains lymphoid tissue in its walls.
(d) is completely covered by peritoneum at its tip.
(e) has a mesentery.

30 (a) **True** Short gastric and left gastro-epiploic arteries are branches of the splenic artery.
(b) **False** The first branch of the superior mesenteric is the inferior pancreaticoduodenal artery.
(c) **True** The anterior vagal trunk is composed mainly of right vagal fibres and the posterior of left vagal fibres.
(d) **True** Veins from the region of the gastro-oesophageal junction, however, drain via the oesophageal veins into the azygos system: this is a region of porto-caval anastomosis.
(e) **True** A posterior gastric ulcer which perforates may allow gastric contents to enter the lesser sac.

31 (a) **True** This is the surface marking for the gall-bladder and it is here that the linea semilunaris crosses the costal margin.
(b) **True** This is the usual origin of the cystic artery.
(c) **False** The common bile duct lies between the leaves of the omentum.
(d) **False** It concentrates bile about 10 times.
(e) **False** Intravenous cholangiography is used to show the duct system. A cholecystogram is used to outline the gall-bladder.

32 (a) **True** The ileocolic artery supplies the terminal ileum, caecum and proximal ascending colon. It usually gives the appendicular artery.
(b) **False** This is a branch of the inferior mesenteric artery.
(c) **False** This is the continuation of the inferior mesenteric artery.
(d) **True** This is the first branch of the artery.
(e) **False** This vessel comes from the hepatic artery.

33 (a) **False** The portal vein does not contain valves.
(b) **False** The union is between the splenic and superior mesenteric veins.
(c) **False** It is the superior mesenteric vessels which are related to the uncinate process of the pancreas.
(d) **True** Consider the boundaries of the opening into the lesser sac. In front, covered with peritoneum is the portal vein: behind, also covered with peritoneum, is the IVC.
(e) **False** The renal veins drain to the IVC.

34 (a) **False** The appendix arises from the posteromedial surface of the caecum. In the child, before the caecum is fully grown, the appendix arises from the tip of a conical caecum.
(b) **False** The appendicular artery arises from a branch of the superior mesenteric artery: either the ileocolic artery itself or its posterior caecal branch.
(c) **True** Collections of lymphoid tissue are a feature of the wall of the appendix.
(d) **True** When there is tissue oedema during appendicitis the peritoneal covering does not allow expansion (compare with the peritoneal covering of the gall-bladder in cholecystitis). The blood vessels become obstructed and gangrene of the tip occurs.
(e) **True** It also often has a peritoneal fold in front of the mesentery, the bloodless fold of Treves.

35 Use the straight abdominal radiograph below to decide which of the following statements are true and which are false.

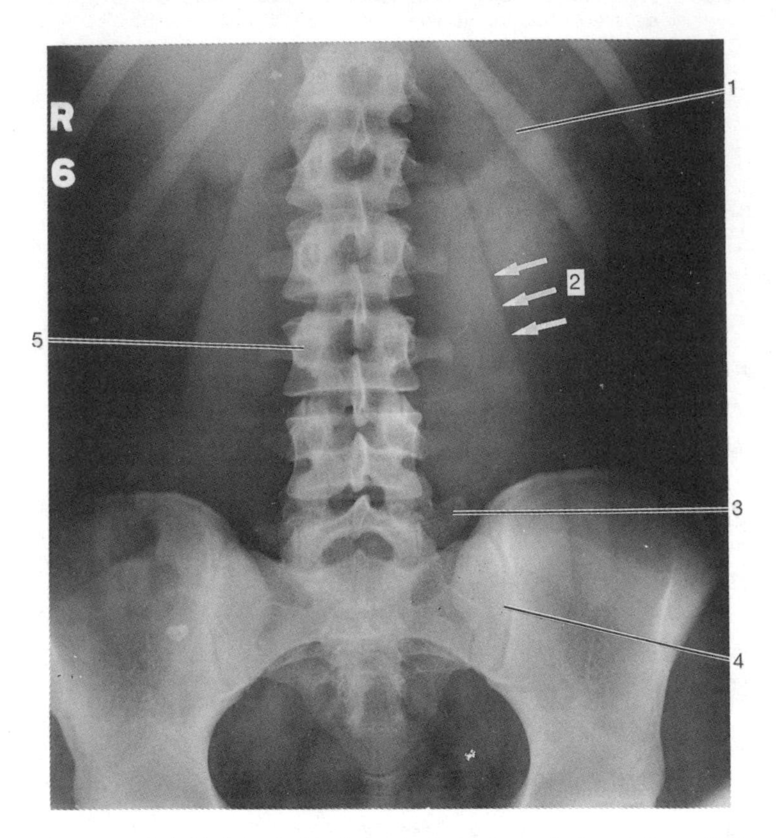

Fig. 2.1 Straight abdominal radiograph. By kind permission of Churchill Livingstone, Publishers. From *Clinical Anatomy in Action*, Volume 3: Pegington.

(a) Bone '1' is without doubt the left 12th rib.
(b) Line '2', marked with white arrows, is the left ureter.
✓ (c) The bony opacity '3' gives attachment to the iliolumbar ligament.
(d) The joint labelled '4' is a fibrous type of joint.
✓ (e) The opaque bony ring in the vertebral shadow labelled '5' is a pedicle.

36 Posterior relations of the first part of the duodenum include
(a) the neck of the pancreas.
(b) the right gastric artery.
✓ (c) the common bile duct.
(d) the accessory pancreatic duct.
(e) the opening into the lesser sac.

35 (a) **False** Although it may look this way at first, count the number of lumbar transverse processes from the one labelled '3' upwards. There are six! It could mean that there are six lumbar vertebrae in this film or that the upper vertebra is T12 with very small 12th ribs. In fact you can just make out these ribs if you look carefully. This question, although not altogether fair, should alert you to get into the habit of counting lumbar vertebrae and looking at them for abnormalities in all abdominal films.

(b) **False** The ureter is not seen on a plain film: the line is the 'psoas' shadow and represents the edge of the psoas muscle.

(c) **True** This is the transverse process of L5 – the largest of the lumbar transverse processes.

(d) **False** The sacro-iliac joint is a synovial joint.

(e) **True** On the anteroposterior film the pedicles of lumbar vertebrae are seen end on in this way.

36 (a) **False** Only the inferior surface of the first part is related to the pancreas.

(b) **False** This artery, often a leash of vessels, runs from the hepatic artery to the distal end of the lesser curvature of the stomach.

(c) **True** The common bile duct runs behind the first part of the duodenum with the gastroduodenal artery on its right.

(d) **False** The accessory pancreatic duct enters the second part of the duodenum, proximal to the opening of the main pancreatic and common bile ducts.

(e) **False** The opening lies above the first part of the duodenum.

37 The inferior mesenteric artery
(a) is related to the duodenum.
(b) supplies the caecum.
(c) has a middle colic artery as one of its branches.
(d) supplies the sigmoid colon.
(e) continues into the pelvis as the superior rectal artery.

38 Use the following diagram of the inferior surface of the liver to answer the question.

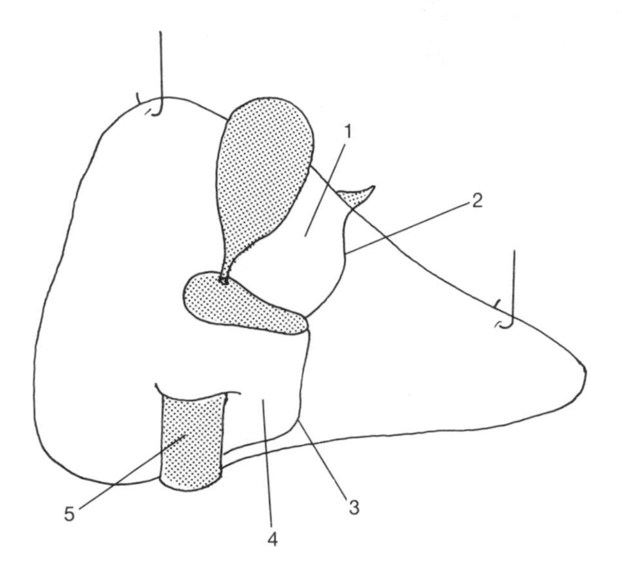

Fig. 2.2 Inferior surface of the liver.

(a) The section of liver labelled 1' is the caudate lobe.
(b) The fibrous cord '2' is a remnant of an umbilical artery.
(c) The fibrous cord '3' is the ligamentum venosum.
(d) The section of liver marked '4' is normally located in the roof of the lesser sac (omental bursa).
(e) The large venous structure '5' receives the hepatic veins.

37 (a) **True** The inferior mesenteric artery arises from the aorta behind the third part of the duodenum.
(b) **False** The artery supplies embryological midgut territory, which starts at the left third of the transverse colon.
(c) **False** This is a branch of the superior mesenteric artery.
(d) **True** Two or three sigmoid branches are usually present.
(e) **True** The vessel continues into the pelvis as the superior rectal artery.

38 (a) **False** This is the quadrate lobe: label '4' is the caudate lobe.
(b) **False** The ligamentum teres is a remnant of the left umbilical vein, which drained into the left branch of the portal vein in the fetal liver.
(c) **True** As the ductus venosus in fetal life, it connected the left branch of the portal vein to the common hepatic vein (later to become the proximal inferior vena cava).
(d) **True** The caudate lobe forms part of the roof of the lesser sac and protrudes into its cavity.
(e) **True** The structure is the IVC.

39 The common bile duct
- (a) lies in the free edge of the lesser omentum.
- (b) is usually joined by the accessory pancreatic duct.
- (c) is related to the gastroduodenal artery.
- (d) is surrounded at its termination by a sphincter.
- (e) is formed by the union of the right and left hepatic ducts.

40 The skin of the scrotum is supplied by
- (a) the ilioinguinal nerve.
- (b) the posterior femoral cutaneous nerve.
- (c) both first lumbar and third sacral segments of the spinal cord.
- (d) branches of the pudendal nerve.
- (e) branches of the obturator nerve.

41 An indirect inguinal hernia
- (a) is only found in females.
- (b) passes down a patent (open) processus vaginalis.
- (c) is a protrusion through the conjoint tendon.
- (d) has its neck lateral to Hesselbach's triangle.
- (e) has the inferior epigastric artery on the medial side of its neck.

39 (a) **True** The hepatic artery lies on its left and the portal vein behind. Although this is the usual arrangement, variations are sometimes found.

(b) **False** The duct is usually joined by the main pancreatic duct to form the hepatopancreatic ampulla.

(c) **True** This vessel usually lies on its left-hand side.

(d) **True** This is the sphincter of Oddi and is usually composed of circular muscle around the termination of the CBD, the pancreatic duct and the ampulla.

(e) **False** These two ducts form the common hepatic duct.

40 (a) **True** This nerve supplies much of the skin of the front of the scrotum.

(b) **True** Branches of this nerve supply skin on the back of the scrotum.

(c) **True** The anterior third of the scrotal skin is supplied by L1 whereas the posterior two-thirds are supplied by S3. The ventral axial line passes between these two segments.

(d) **True** Scrotal branches of the nerve supply scrotal skin.

(e) **False** The obturator nerve takes no part in the supply of the scrotal skin. The only other nerve to supply scrotal skin apart from those mentioned in (a), (b) and (d) is the genital branch of the genitofemoral nerve (Gray's Anatomy).
This is denied by Last (Regional Applied Anatomy).

41 (a) **False** Although more common in males, such a hernia does occur in females.

(b) **True** The hernia is sometimes found in infants or in young adults. In the latter cases it is often related to a strain while lifting a heavy load.

(c) **False** It is a direct inguinal hernia that pushes through a weakened posterior inguinal canal wall or conjoint tendon.

(d) **True** The hernia enters the patent processus at the deep ring lateral to the triangle, the boundaries of which are the inferior epigastric artery, inguinal ligament and lateral edge of the rectus. The triangle can only be seen from the abdominal side of the lower abdominal wall.

(e) **True** The artery lies to the medial side of the neck of the hernia: it lies lateral to a direct inguinal hernia.

42 Posterior relations of the left kidney include
(a) the twelfth rib.
(b) the tail of the pancreas.
(c) the genitofemoral nerve.
(d) the ilioinguinal nerve.
(e) the subcostal nerve and vessels.

43 Concerning the kidneys:
(a) pararenal fat (Zuckerkandl) lies between the capsule and the renal fascia.
(b) they are retroperitoneal structures.
(c) the left kidney usually lies at a lower level than the right.
(d) the lateral arcuate ligament is a posterior relation on both sides.
(e) pleura is a posterior relationship of both kidneys.

44 The abdominal part of the oesophagus
(a) enters the abdomen at the level of T12.
(b) enters the abdomen between fibres of the right crus.
(c) gains its blood supply from a branch of the coeliac trunk.
(d) is held partly in place by the phreno-oesophageal ligament.
(e) lies behind the left lobe of the liver.

45 The abdominal part of the left sympathetic trunk
(a) becomes continuous with the thoracic part behind the lateral arcuate ligament.
(b) is a retroperitoneal structure.
(c) passes behind the common iliac artery.
(d) is closely related to the abdominal aorta.
(e) receives preganglionic sympathetic fibres from the 5th lumbar spinal nerve.

42 (a) **True** During some operations on the kidney by the lumbar route the 12th rib is removed.

(b) **False** The pancreas runs in front of the kidney.

(c) **False** The genitofemoral nerve is located on the front of the psoas major at a lower level.

(d) **True** The iliohypogastric nerve is also a posterior relation of the kidney.

(e) **True** The subcostal neurovascular bundle emerges from behind the lateral arcuate ligament and travels behind the kidney.

43 (a) **False** Pararenal fat (Zuckerkandl) is found BEHIND the renal fascia: Perirenal fat (Gerota) lies between the capsule and renal fascia.

(b) **True** Both kidneys are found in the retroperitoneal region.

(c) **False** The left kidney lies higher than the right.

(d) **True** Posterior muscular relationships include the psoas, quadratus lumborum, diaphragm, and transversus abdominis.

(e) **True** During the lumbar approach to the kidneys care must be taken not to open the pleura.

44 (a) **False** The oesophagus enters the abdomen at the level of T10: the IVC pierces the diaphragm at T8 and the aorta at T12.

(b) **True** The fibres of the right crus run on both sides of the oesophagus, and these form one mechanism for preventing gastro-oesophageal reflux.

(c) **True** The left gastric artery supplies oesophageal branches.

(d) **True** This cone-shaped ligament is attached at the edge of the oesophageal opening of the diaphragm below, and the oesophagus a few centimeters higher up.

(e) **True** The oesophagus leaves a groove at the back of the left lobe of a formalin-hardened liver.

45 (a) **False** The sympathetic trunk runs behind the medial arcuate ligament.

(b) **True** The trunks lie behind the peritoneum of the posterior abdominal wall.

(c) **True** Below this level it joins the right-sided trunk to form the ganglion impar at the level of the coccyx.

(d) **True** The trunk lies to the left of the aorta and the lateral aortic lymph nodes. The right trunk is overlapped by the IVC.

(e) **False** The preganglionic outflow from the cord only reaches L2 or L3.

46 Use the following photograph of a CT of the abdomen to answer the question. **Contrast has been introduced into the arteries.**

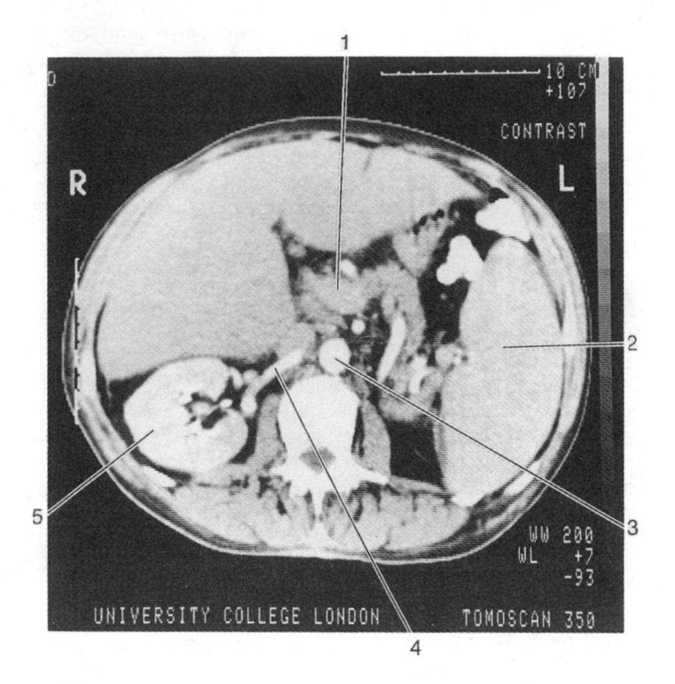

Fig. 2.3 CT of abdomen. By kind permission of Churchill Livingstone, Publishers. From *Clinical Anatomy in Action*, Volume 3: Pegington.

(a) Label '1' is part of the gut.
(b) Structure '2' is the spleen.
(c) Structure '3' is the abdominal aorta.
(d) Vessel '4' is the hepatic artery.
(e) Structure '5' is a suprarenal gland.

47 Concerning the surface anatomy and examination of the normal spleen:
(a) its inferior border may normally be palpated in the epigastrium.
(b) it lies deep to the 9th, 10th and 11th left ribs.
(c) its anterior limit is the anterior axillary line.
(d) a notch is usually palpable.
(e) an area of colonic resonance running across the spleen will be found during percussion of the abdomen.

46 (a) **False** The organ lying transversely across the abdomen in front of the aorta is the pancreas.
(b) **True** Note the close relationship of the tail of the pancreas to the splenic hilum.
(c) **True** The aorta contains contrast in this CT.
(d) **False** This is the right renal artery.
(e) **False** This is the right kidney: each suprarenal gland looks like an inverted 'V' and of course is much smaller than object '5'.

47 (a) **False** The normal spleen is not palpable.
(b) **True** Posteriorly it lies 4 cm from the mid- dorsal line and its long axis lies along the 10th rib.
(c) **False** The anterior limit is the mid-axillary line.
(d) **False** A notch is usually found on the superior border, but it only becomes palpable with marked enlargement of the organ.
(e) **False** The area of splenic dullness is found over the 9th, 10th and 11th left ribs as far as the mid-axillary line. When the spleen is enlarged, however, the area is more extensive, and the organ extends and pushes the colon towards the right iliac fossa. An enlarged left kidney, on the other hand, also presents with a mass dull to percussion, but because it is a retroperitoneal structure the colon is simply pushed forwards. The result is a strip of resonance across the dull mass.

48 A Meckel's diverticulum
  (a) projects from the mesenteric border of the ileum.
  (b) is found on average 1 metre above the ileocaecal valve.
  (c) is a persistent urachus.
  (d) is found in 30% of subjects.
  (e) may be lined with mucous membrane similar to that of the stomach.

49 Concerning the superficial inguinal ring:
  (a) it is a defect in the external oblique aponeurosis.
  (b) its medial crus is attached to the pubic tubercle.
  (c) fibres from its edges are continued over the spermatic cord as the cremaster muscle.
  (d) its surface marking is the mid-inguinal point.
  (e) it is strengthened by intercrural fibres.

50 The abdominal part of the right ureter
  (a) travels on the right psoas muscle.
  (b) crosses in front of the right genitofemoral nerve.
  (c) is related to the gonadal vessels.
  (d) runs behind the duodenum.
  (e) runs behind the middle colic artery.

48 (a) **False** The diverticulum projects from the antimesenteric border of the ileum.
(b) **True** This is the site of the apex of the embryonic midgut loop.
(c) **False** The median umbilical ligament is a remnant of the urachus: a Meckel's diverticulum is a remnant of the vitelline or yolk duct.
(d) **False** A Meckel's diverticulum is said to occur in about 3% of individuals.
(e) **True** The mucosa may even secrete acid.

49 (a) **True** The external ring is an opening in the external oblique aponeurosis just above the crest of the pubis.
(b) **False** It is the lateral crus which is attached to the tubercle, the medial crus is attached to the symphysis.
(c) **False** The cremasteric fascia and muscle are derived from the internal oblique and transversus: the external oblique is continued over the cord as the external spermatic fascia.
(d) **False** The mid-inguinal point may be used as a surface mark for the deep inguinal ring which lies about 1.25 cm above the inguinal ligament.
(e) **True** Intercrural fibres are found at the apex of the 'ring' binding the crura together.

50 (a) **True** The ureter may be found on the psoas muscle.
(b) **True** The genitofemoral nerve is also found running on the front of the psoas.
(c) **True** The gonadal vessels cross in front of the ureter on the psoas.
(d) **True** The right ureter runs behind the second part of the duodenum from the pelvis.
(e) **False** The right ureter crosses behind the right colic and ileocolic vessels.

Pelvis and Perineum

51 The deep perineal pouch in the male
- (a) contains the bulbourethral glands.
- (b) transmits the membranous urethra.
- (c) contains the ischiocavernosus muscles.
- (d) contains the greater vestibular glands.
- (e) contains the sphincter urethrae

52 Structures found within the peritoneal fold called the broad ligament include
- (a) the uterine tubes.
- (b) the suspensory ligament of the ovary.
- (c) part of the round ligament of the uterus.
- (d) the ovarian ligament (round ligament of the ovary).
- (e) branches of the uterine artery.

53 Concerning the internal female genital organs:
- (a) the commonest position of the uterus is retroverted.
- (b) the posterior fornix of the vagina is related to the pouch of Douglas (recto-uterine pouch).
- (c) the ovary is suspended from the anterior leaf of the broad ligament.
- (d) the uterine artery is a branch of the abdominal aorta.
- (e) the round ligament of the uterus is a remnant of the gubernaculum.

54 Anal mucous membrane above the level of the pectinate line
- (a) has a lymph drainage into the superficial inguinal nodes.
- (b) has a venous drainage to the portal system.
- (c) receives sensory supply from the pudendal nerve.
- (d) receives its arterial supply from branches of the inferior mesenteric artery.
- (e) is characterised by anal columns.

51 (a) **True** The ducts of these glands, however, pierce the perineal membrane before entering the spongiose urethra.
(b) **True** This is the least dilatable and shortest part of the urethra.
(c) **False** This covers the crus and is found in the superficial pouch.
(d) **False** These glands are only found in the female.
(e) **True** This muscle and the deep transverse perinei form the muscular contents of the deep pouch.

52 (a) **True** The tube lies in the free border of the ligament.
(b) **True** This 'ligament' is composed of fibrous tissue surrounding the ovarian vessels.
(c) **True** The distal section of the ligament runs through the inguinal canal.
(d) **True** This connects the ovary to the uterus and is a remnant of the proximal end of the gubernaculum.
(e) **True** The artery is closely associated with the transverse cervical ligament (Mackenrodt).

53 (a) **False** The normal uterus is anteverted and anteflexed.
(b) **True** A 'mass' in the pouch may therefore be palpated by examining fingers placed in this fornix.
(c) **False** The mesovarium is continuous with the posterior leaf.
(d) **False** The artery is a branch of the internal iliac artery.
(e) **True** The round ligaments of the ovary and uterus are both remnants of the gubernaculum.

54 (a) **False** Lymphatic drainage passes to para-aortic nodes.
(b) **True** The superior rectal veins drain towards the portal system.
(c) **False** The mucosa is only sensitive to stretch in this region and these sensory fibres follow autonomic pathways back to the spinal cord.
(d) **True** The superior rectal artery is a branch of the inferior mesenteric artery.
(e) **True** These so-called columns of Morgagni are mucosal folds containing submucous rectal veins.

55 The prostate
(a) is separated from the rectum by a fascial septum.
(b) is pierced by the ducts of the seminal vesicles.
(c) is related in front to an extraperitoneal space (space of Retzius).
(d) possesses a median lobe.
(e) has a rich venous plexus associated with its fascial sheath.

56 The right ovary
(a) is attached to the broad ligament by the suspensory ligament of the ovary.
(b) has a derivative of the gubernaculum attached to its inferior pole.
(c) is closely related to the obturator nerve.
(d) has a lymphatic drainage to the para-aortic nodes.
(e) has a venous drainage to the inferior vena cava.

57 The superficial perineal pouch in the male
(a) is bounded below by the membranous layer of superficial fascia (Colles).
(b) is bounded above by the dartos muscle.
(c) contains the bulb of the penis.
(d) is open posteriorly.
(e) communicates freely with the ischiorectal fossa.

58 The pudendal nerve
(a) is a branch of the sacral plexus.
(b) leaves the pelvic cavity through the lesser sciatic foramen.
(c) enters the perineum by piercing the levator ani muscle.
(d) supplies the superficial perineal muscles in the male.
(e) gives branches which supply scrotal skin.

59 The external anal sphincter
(a) is the lowermost part of the circular muscle of the intestine.
(b) is composed of three parts.
(c) is supplied by the inferior rectal nerve.
(d) is involuntary (smooth muscle).
(e) is attached to the perineal body.

55 (a) **True** This is the rectovesical fascia of Denonvilliers.
(b) **False** Each duct joins its corresponding ductus deferens.
(c) **True** The medial puboprostatic ligaments form the floor of this space.
(d) **True** This is usually defined as that part of the gland between the ejaculatory ducts and the urethra.
(e) **True** This is especially rich on each side of the gland.

56 (a) **False** The ovary is attached by the mesovarium.
(b) **True** The ovarian ligament is attached to its inferior pole (Q.53 (e))
(c) **True** Ovarian problems may give rise to pain referred along the nerve.
(d) **True** The lymphatics follow the pathway of the ovarian vessels.
(e) **True** The left ovarian vein drains into the left renal vein.

57 (a) **True** The fascia is continuous over the lower abdominal wall with the membranous layer of superficial fascia.
(b) **False** It is bounded above by the perineal membrane.
(c) **True** It also contains the crura of the penis.
(d) **False** Behind, Colles' fascia is attached to the posterior edge of the perineal membrane.
(e) **False** The space communicates with a space deep to the membranous layer of abdominal wall fascia. Urine leaking from a ruptured urethra tracks from the perineum into the lower abdominal wall by this route.

58 (a) **True** Its root value is S.2, S.3 and S.4.
(b) **False** It leaves the pelvis through the greater sciatic foramen.
(c) **False** It enters the perineum through the lesser sciatic foramen.
(d) **True** The perineal branch of the nerve supplies these muscles.
(e) **True** The perineal branch also gives scrotal branches.

59 (a) **False** The internal sphincter is part of the circular muscle.
(b) **True** These are called subcutaneous, superficial and deep parts.
(c) **True** This is a branch of the pudendal nerve.
(d) **False** It is the internal sphincter which is involuntary.
(e) **True** The main attachment to this structure is through the superficial part of the sphincter.

60 Use the following CT of a female pelvis to answer the question.

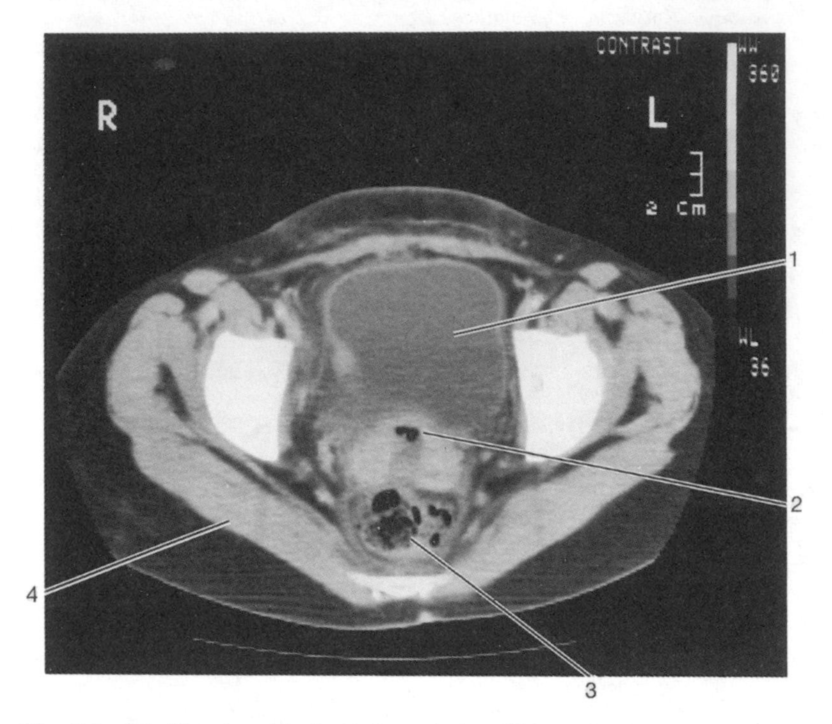

Fig. 3.1 CT of female pelvis. By kind permission of Churchill Livingstone, Publishers. From *Clinical Anatomy in Action*, Volume 3: Pegington.

(a) The section is made at the level of S1.
(b) Structure '1' is the uterus.
(c) Structure' 2' is the aorta.
(d) Structure '3' is part of the gastrointestinal tract.
(e) Muscle '4' is an extensor of the leg at the hip joint.

61 The ischiorectal fossa
(a) is bounded superiorly by the levator ani muscle.
(b) is bounded laterally by the pubic ramus.
(c) is bounded medially by the rectum.
(d) has the pudendal canal running along its lateral wall.
(e) is bounded above by the lunate fascia.

60 (a) **False** The level is below the lumbosacral canal at the level of the lower sacrum or coccyx.
(b) **False** Structure '1' is the bladder.
(c) **False** Structure '2' is the cervix.
(d) **True** This is the lower rectum and contains air.
(e) **True** This is the gluteus maximus.

61 (a) **True** The upper edge of the fossa is formed by the meeting of the levator ani and obturator internus.
(b) **False** The lateral boundary is the fascia covering the obturator internus.
(c) **False** The rectum is a pelvic structure, it is the anal canal which lies medial to the fossa.
(d) **False** This canal contains the pudendal neurovascular bundle.
(e) **True** This fascia extends upwards from the pudendal canal fascia to the roof. It is said to be the true roof and lateral wall of the fossa.

62 The uterine tube
(a) is narrowest at the ampulla.
(b) is developed from the mesonephric duct.
(c) receives blood from both ovarian and uterine arteries.
(d) is about 10 cm long in the adult.
(e) is usually the place where fertilization of the ovum takes place.

63 The rectum
(a) has a mesentery.
(b) has appendices epiploicae.
(c) has muscle in its wall supplied by parasympathetic motor fibres.
(d) is supplied with blood by the inferior rectal artery.
(e) is closely related to the inferior hypogastric plexuses.

64 Concerning the urinary bladder
(a) its sphincter vesicae is supplied by sympathetic motor fibres.
(b) it is connected to a remnant of the urachus.
(c) it lies against the fascia of the levator ani.
(d) its detrusor muscle is supplied by sympathetic motor fibres.
(e) it lies above the level of the pelvic inlet in the new-born infant.

65 The female ureter
(a) is related to the ovarian fossa.
(b) is related to the obturator neurovascular bundle.
(c) runs above the uterine artery.
(d) runs behind the internal iliac artery.
(e) runs through the bladder wall obliquely.

62 (a) **False** The isthmus is the narrowest part of the tube and has the thickest wall.
(b) **False** The tubes develop from the paramesonephric ducts.
(c) **True** The two arteries anastomose somewhere along the tube: the uterine artery usually supplying about the medial two-thirds of the tube.
(d) **True** It is described as having a fimbriated end, infundibulum, ampulla, isthmus, and uterine section.
(e) **True** Fertilization usually takes place in the ampulla.

63 (a) **False** Peritoneum only covers the front and sides of the upper third and the front of the middle third of the rectum.
(b) **False** These are found on the surface of the colon. The rectum has no sacculations and no mesentery.
(c) **True** The pelvic splanchnic nerves supply the rectal musculature.
(d) **False** The inferior 'rectal' artery supplies the lower anal canal.
(e) **True** The plexuses run on the sides of the rectum. Damage to them during resection of the rectum may cause post-operative impotence in the male.

64 (a) **True** The parasympathetic nervi erigentes (S.2, 3 and 4), on the other hand, supply inhibitory fibres to this muscle.
(b) **True** The remnant of the urachus, the median umbilical ligament, runs from the apex of the bladder to the anterior abdominal wall.
(c) **True** In both sexes the inferolateral surfaces are related to these muscles and to the obturator internus.
(d) **False** The detrusor is supplied by parasympathetic fibres (S.2, 3 and 4). The sympathetic fibres are inhibitory.
(e) **True** At birth the bladder is an abdominal rather than a pelvic organ.

65 (a) **True** Boundaries of the fossa include the ureter, internal iliac artery and the obliterated umbilical artery.
(b) **True** The ureter runs medial to these structures.
(c) **False** The ureter runs below the artery.
(d) **False** It runs anterior to this vessel.
(e) **True** This obliquity is an important factor in prevention of regurgitation of urine from the bladder to the ureters.

66 Use the CT of the male pelvis to answer the question. **Contrast has been introduced into the urinary tract.**

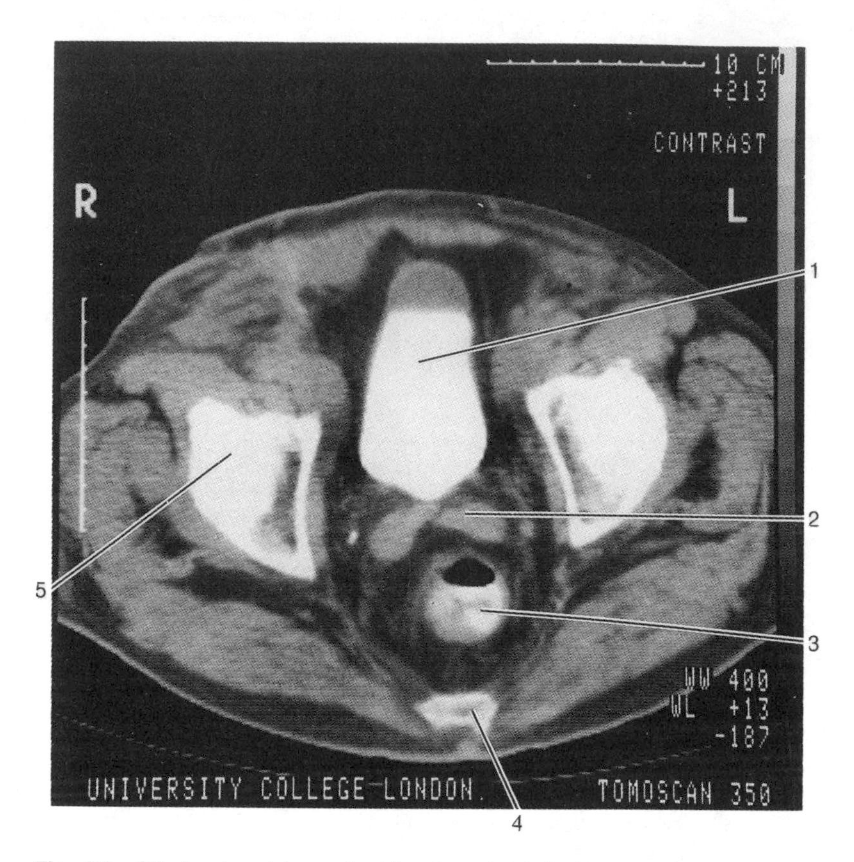

Fig. 3.2 CT of male pelvis: contrast in urinary tract. By kind permission of Churchill Livingstone, Publishers. From *Clinical Anatomy in Action*, Volume 3: Pegington.

(a) Structure '1' normally contains urine.
(b) Structure '2' is the ductus (vas) deferens.
(c) Structure '3' is the rectum.
(d) Structure '4' lies below the level of S1.
(e) Structure '5' is the femur.

67 Each seminal vesicle
(a) is palpable by digital rectal examination.
(b) consists of a single coiled tube.
(c) lies medial to its corresponding ductus (vas) deferens.
(d) is in direct contact with the bladder.
(e) is separated from the rectum by rectovesical fascia (fascia of Denonvilliers).

66 (a) **True** Structure '1' is the bladder.
(b) **False** The large structures behind the bladder are the seminal vesicles.
(c) **True** This contains air (black areas).
(d) **True** It is probably the lowest part of the sacrum or the coccyx.
(e) **False** This is part of the pelvis and is best worked out with the aid of Fig. 3.3 below.

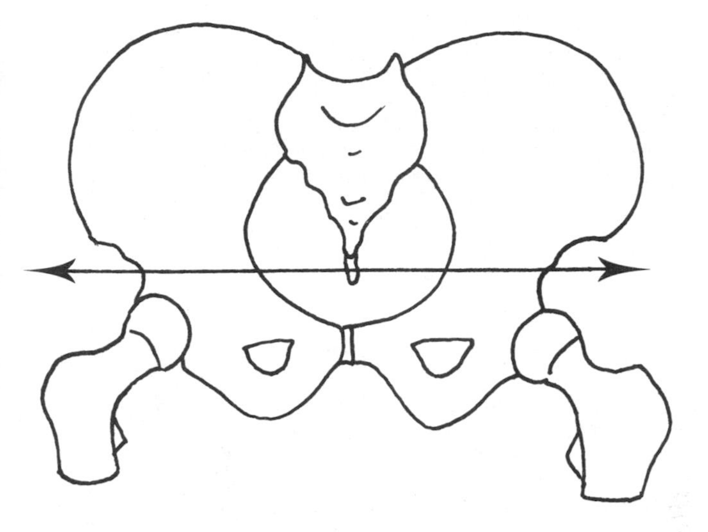

Fig. 3.3 Level of section in Fig. 3.2.

67 (a) **True** The vesicles may be palpated at the tip of the examining finger.
(b) **True** The tube is about 10–15 cm long.
(c) **False** Each vesicle lies lateral to the corresponding ampulla of the ductus (vas) deferens.
(d) **True** The seminal vesicles lie against the posterior surface of the bladder.
(e) **True** The fascia lies behind the seminal vesicles and between the posterior surface of the bladder and rectum.

68 Each greater vestibular gland
- (a) is a homologue of the seminal vesicle.
- (b) opens by means of 15 or more ducts.
- (c) lies close to the bulb of the vestibule.
- (d) is located deep to the urogenital diaphragm.
- (e) is the sole source of lubrication for the vagina.

69 Concerning the vagina:
- (a) its posterior wall is longer than its anterior wall.
- (b) its posterior fornix is deeper than the anterior fornix.
- (c) it receives a blood supply from the uterine arteries.
- (d) it is closely related to the urethra.
- (e) its posterior wall is completely covered with peritoneum.

70 Concerning the diameters and shape of the pelvis of a typical adult, white, female subject:
- (a) the obstetric conjugate is measured between the sacral promontory and the lower edge of the symphysis pubis.
- (b) the anteroposterior diameter of the inlet is usually shorter than its transverse diameter.
- (c) the distance between the ischial spines is usually the widest part of the pelvic cavity.
- (d) the commonest basic type is known as a platypelloid pelvis.
- (e) the plane of greatest diameter lies at the level of S2–S3.

71 Muscles of the urogenital diaphragm include
- (a) levator ani.
- (b) sphincter urethrae.
- (c) sphincter vesicae.
- (d) ischiocavernosus.
- (e) obturator internus.

68 (a) **False** They are homologues of the bulbourethral glands.
(b) **False** There is one duct for each gland: it opens in the groove between the labium minus and the hymen.
(c) **True** Each gland is overlapped by the posterior end of the bulb.
(d) **False** Although the bulbourethral glands in the male lie deep to the membrane, in the deep perineal pouch, the greater vestibular glands lie more superficial in the perineum.
(e) **False** The vagina is mostly lubricated by secretions from glands in the cervix.

69 (a) **True** The average length is 7.5 cm for the anterior wall and 9 cm for the posterior wall.
(b) **True** Examination of the posterior fornix is important because the pouch of Douglas may be palpated through its wall.
(c) **True** Arterial supply comes from vaginal, uterine, middle rectal and internal pudendal arteries.
(d) **True** The urethra is actually embedded in the anterior wall.
(e) **False** Only the upper quarter is covered by peritoneum of the pouch of Douglas.

70 (a) **False** This is the diagonal (oblique) conjugate. It is measured during vaginal examination and may be used to estimate the true or obstetric conjugate from promontory to the upper border of the symphysis (average 11 cm).
(b) **True** The transverse diameter is of the order of 13 cm.
(c) **False** This is usually the smallest diameter of the pelvic cavity, and lies at the plane of least dimensions.
(d) **False** Platypelloid pelves are found in only 2.6% of adult white female pelves, anthropoid in 23.5%, android in 32.5% and gynaecoid in 41.4%.
(e) **True** The pelvis is most capacious at this level.

71 (a) **False** The muscles belong to the pelvic diaphragm.
(b) **True** The only other muscle fibres which form part of the urogenital diaphragm are derived from the deep transverse perinei.
(c) **False** The sphincter vesicae is proximal to the diaphragm around the internal urethral orifice.
(d) **False** The ischiocavernosus is located in the superficial perineal pouch.
(e) **False** Obturator internus is a muscle of the pelvic wall.

72 In the penis
  (a) the deep artery supplies the cavernous tissue of the corpus cavernosum.
  (b) the deep dorsal vein drains to the prostatic plexus of veins.
  (c) the dorsal artery runs lateral to the dorsal nerve.
  (d) the crus penis is attached to the ischiopubic ramus.
  (e) the narrowest part of the urethra is found.

73 Levator ani
  (a) is a muscle of the pelvic diaphragm.
  (b) inserts partly into the perineal body.
  (c) arises from the obturator membrane.
  (d) is covered on both sides by fascia.
  (e) is supplied by the obturator nerve.

74 Obturator internus
  (a) arises in part from the obturator membrane.
  (b) is covered on its internal surface by strong fascia.
  (c) leaves the pelvis through the greater sciatic foramen.
  (d) inserts into the greater trochanter of the femur.
  (e) is supplied by the 4th sacral spinal nerve.

75 Concerning common differences between adult male and female pelves:
  (a) the subpubic angle is usually greater in the male than in the female.
  (b) in the male the diameter of the acetabulum is much less than the distance from the anterior acetabular rim to the symphysis pubis.
  (c) the ischial spines tend to be further apart in the female than in the male pelvis.
  (d) the sacrum is less curved in the female.
  (e) the greater sciatic notch tends to be wider in the male than in the female.

72 (a) **True** The deep artery of the penis, a branch of the internal pudendal artery enters the crus of the penis.
(b) **True** The vein also communicates with the internal pudendal veins.
(c) **False** The artery runs between the dorsal nerve and the deep dorsal vein.
(d) **True** The crura are the posterior ends of the corpora cavernosa and each gains anchorage to the bone in front of the ischial tuberosity.
(e) **True** The external urethral orifice is the narrowest part of the urethra. The next narrowest part is the membranous urethra.

73 (a) **True** The diaphragm consists of this muscle and coccygeus.
(b) **True** The fibres which constitute either levator prostatae (male) or pubovaginalis (female) insert into the body.
(c) **False** The muscle arises from the fascia covering the medial surface of the obturator internus.
(d) **True** The superior fascia is part of the parietal pelvic fascia and the inferior fascia a boundary of the ischiorectal fossa.
(e) **False** The muscle is supplied by a branch of the fourth sacral spinal nerve and a branch of the inferior rectal nerve.

74 (a) **True** The muscle arises from the bone surrounding the obturator foramen and the obturator membrane. Above, it arises from a tendinous arch which spans the obturator canal.
(b) **True** The fascia is thickened as the 'tendinous arch of the levator ani' along or near the origin of this muscle from the fascia.
(c) **False** The muscle leaves through the lesser sciatic foramen.
(d) **True** The insertion is into the medial surface of the greater trochanter just above the trochanteric fossa.
(e) **False** The nerve to obturator internus is a branch of the sacral plexus (L5 S1). Levator ani is supplied by S4 (Q.73 (e)).

75 (a) **False** The subpubic angle is greater in the female.
(b) **False** The acetabulum is absolutely larger than in the female and its diameter is about equal to the distance between its anterior rim and the symphysis. In the female the diameter is much less than this distance, reflecting both the small size of the acetabulum and the wide pelvis.
(c) **True** In the male they are closer together and turned in.
(d) **True** It is particularly flattened in the region of S2 and S3. The male sacrum is more evenly curved and tends to be long and narrow.
(e) **False** The notch is much wider in the female.

Note: The student should expect to find considerable overlap in characteristics between the sexes, and not all features will indicate a 'male' or 'female' pelvis in every case.

Head and Neck

76 The sternocleidomastoid muscle
   (a) has a muscular origin from the clavicle.
   (b) inserts into the temporal bone.
   (c) receives its motor supply from vagal fibres.
   (d) may be used as a surface marker to find a large vein in the neck for cannulation.
   (e) rotates the head so that it looks upwards to the side under test.

77 Branches of the external carotid artery include
   (a) the inferior thyroid artery.
   (b) the vertebral artery.
   (c) the costocervical trunk.
   (d) the occipital artery.
   (e) the facial artery.

78 Branches of the cervical plexus include
   (a) the auriculotemporal nerve.
   (b) the buccal nerve.
   (c) the suprascapular nerves.
   (d) the phrenic nerve.
   (e) the great auricular nerve.

79 Structures which are found in the posterior triangle of the neck include
   (a) the glossopharyngeal nerve.
   (b) the common carotid artery.
   (c) the accessory nerve.
   (d) the levator scapulae muscle.
   (e) the upper trunk of the brachial plexus.

80 The lingual artery
   (a) is a branch of the subclavian artery.
   (b) has its origin from the external carotid at the level of the cricoid cartilage.
   (c) is crossed superficially by the hypoglossal nerve.
   (d) runs through the substance of the submandibular salivary gland.
   (e) runs deep to the hyoglossus muscle.

76 (a) **True** It also arises by a tendinous head from the manubrium.
(b) **True** The muscle inserts into the mastoid process and along the superior nuchal line.
(c) **False** The motor supply is from the spinal part of the accessory nerve.
(d) **True** The triangle made by the two origins of the muscle and the clavicle is a surface mark for the internal jugular vein.
(e) **False** The head is rotated to the opposite side.

77 (a) **False** This vessel is a branch of the thyrocervical trunk of the subclavian artery.
(b) **False** The vessel is a branch of the subclavian artery.
(c) **False** This is also a branch of the subclavian artery: other branches of the subclavian are the internal thoracic and dorsal scapular arteries.
(d) **True** At its origin this vessel is crossed superficially by the hypoglossal nerve.
(e) **True** The vessel arises at the level of the hyoid bone.

78 (a) **False** This is a branch of the mandibular nerve.
(b) **False** This is a sensory branch of the mandibular nerve.
(c) **True** These are usually three in number.
(d) **True** The phrenic nerve arises from C3, C4 and C5.
(e) **True** It emerges around the posterior border of the sternomastoid, travels upwards on that muscle and divides into branches on the surface of the parotid.

79 (a) **False** The nerve is located on the stylopharyngeus muscle.
(b) **False** Each common carotid runs behind the sternoclavicular joint and sternocleidomastoid.
(c) **True** The accessory nerve runs obliquely across the triangle supplying sternocleidomastoid and trapezius.
(d) **True** It is found in the floor of the triangle.
(e) **True** All three trunks are located in the posterior triangle.

80 (a) **False** The vessel is a branch of the external carotid.
(b) **False** The vessel lies at the level of the greater cornu of the hyoid.
(c) **True** The nerve crosses a loop in the artery near the hyoid.
(d) **False** It is the facial artery which takes this course.
(e) **True** The vessel runs deep to the muscle to reach the tongue.

81 Structures which pass through the foramen magnum include
(a) the dura.
(b) the vertebral arteries.
(c) the spinal accessory nerves.
(d) the vagi.
(e) the anterior spinal artery.

82 The thyroid gland
(a) is surrounded by a sheath of pretracheal fascia.
(b) has an isthmus covering the 2nd, 3rd and 4th tracheal rings.
(c) receives blood from a branch of the external carotid artery.
(d) receives blood from a branch of the costocervical trunk.
(e) may receive blood from a branch of the aorta.

83 The parotid gland
(a) receives secretomotor fibres which have synapsed in the pterygopalatine ganglion.
(b) contains the retromandibular vein.
(c) contains branches of the seventh cranial nerve.
(d) is surrounded by a dense sheath of fascia.
(e) has a duct which opens opposite the upper second molar tooth.

84 In the neonatal skull
(a) the anterior fontanelle closes at birth.
(b) the mastoid process is long and well developed.
(c) the angle of the mandible is about 90 degrees.
(d) the maxillary air sinus is fully developed.
(e) the tympanic plate is fully formed.

85 The glossopharyngeal nerve
(a) supplies fibres to the carotid sinus.
(b) supplies stylopharyngeus.
(c) supplies taste fibres on the anterior third of the tongue.
(d) sends fibres to the middle ear.
(e) supplies motor fibres to the constrictor muscles of the pharynx.

81 (a) **True** The spinal and cranial dura are continuous through the foramen.
(b) **True** The vessels cross the posterior arch of the atlas, enter the vertebral canal, and ascend through the foramen magnum.
(c) **True** The spinal accessory nerves ascend through the foramen to reach the jugular foramina.
(d) **False** The vagi leave the skull through the jugular foramina.
(e) **True** The anterior spinal arteries, branches of the vertebral arteries, unite in front of the medulla and the common trunk descends through the foramen magnum.

82 (a) **True** The sheath is attached to the thyroid cartilage and by the ligament of Berry to the cricoid. It is these attachments which cause the gland to rise with the larynx during swallowing.
(b) **True** The isthmus is divided in this position during tracheostomy.
(c) **True** The branch is the superior thyroid artery.
(d) **False** The inferior thyroid artery is a branch of the thyrocervical trunk of the subclavian.
(e) **True** Occasionally a thyroidea ima supplies the gland.

83 (a) **False** The synapses for secretomotor fibres lie in the otic ganglion.
(b) **True** The vein lies superficial to the external carotid within the substance of the gland.
(c) **True** The nerve lies superficial to both retromandibular vein and external carotid artery.
(d) **True** Incision of the fascia is occasionally required for relief of pressure during inflammation of the gland.
(e) **True** The duct pierces the buccinator to reach this site.

84 (a) **False** It closes at about the age of 18 months.
(b) **False** The process is not found in the neonatal skull.
(c) **False** The angle of the mandible is more obtuse than this.
(d) **False** None of the paranasal air sinuses are developed at this stage.
(e) **False** The tympanic part of the temporal bone is only found as a tympanic ring, and later grows into the larger plate.

85 (a) **True** It may also receive vagal fibres.
(b) **True** This is the only muscle supplied by the nerve.
(c) **False** It supplies ordinary and special sensation on the posterior third of the tongue.
(d) **True** The tympanic branch (Jackobson's nerve) supplies sensory and parasympathetic fibres to the tympanic plexus.
(e) **False** The nerve supplies only sensory fibres to the pharyngeal plexus. Motor fibres are supplied by the cranial part of the accessory nerve (through the pharyngeal branch of the vagus).

86 Use the following diagram to answer this question. It illustrates the INNER surface of the mandible.

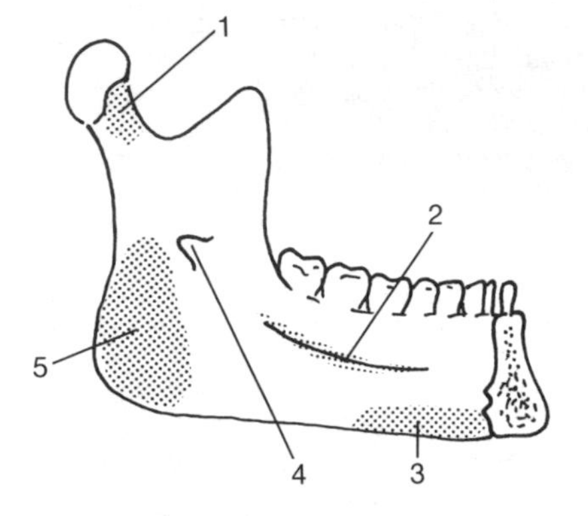

Fig. 4.1 Diagram of the inner surface of the mandible.

(a) The muscle attached to area '1' gains its motor supply from the facial nerve.
(b) The muscle attached to line '2' receives its motor supply from a branch of the inferior alveolar nerve.
(c) The muscle belly arising from fossa '3' is supplied by a branch of the facial nerve.
(d) The bony projection '4' gives attachment to the stylomandibular ligament.
(e) the medial pterygoid is attached at area '5'.

87 The lateral pterygoid muscle
(a) is supplied by a branch of the maxillary nerve.
(b) inserts into the articular disc of the TMJ.
(c) elevates the mandible (closes the mouth).
(d) forms part of the bed of the palatine tonsil.
(e) arises from the medial pterygoid plate.

86 (a) **False** The muscle concerned is the lateral pterygoid and it is supplied by the mandibular nerve.
(b) **True** The muscle is the mylohyoid.
(c) **False** The muscle is the anterior belly of digastric and is supplied by a branch from the nerve to mylohyoid.
(d) **False** The sphenomandibular ligament is attached to the bony process, the lingula.
(e) **True** On the outer side of the mandible, the masseter muscle insertion occupies a similar position.

87 (a) **False** The muscles of mastication are supplied by the mandibular nerve.
(b) **True** The muscle also inserts into a pit on the neck of the mandible and into the capsule of the TMJ.
(c) **False** The lateral pterygoid pulls the head of the mandible forwards and opens the mouth.
(d) **False** The bed of the tonsil is formed by the superior constrictor of the pharynx.
(e) **False** The muscle arises from the lateral pterygoid plate. An upper head arises from the greater wing of the sphenoid.

88 Use the following radiograph of the skull to answer the question.

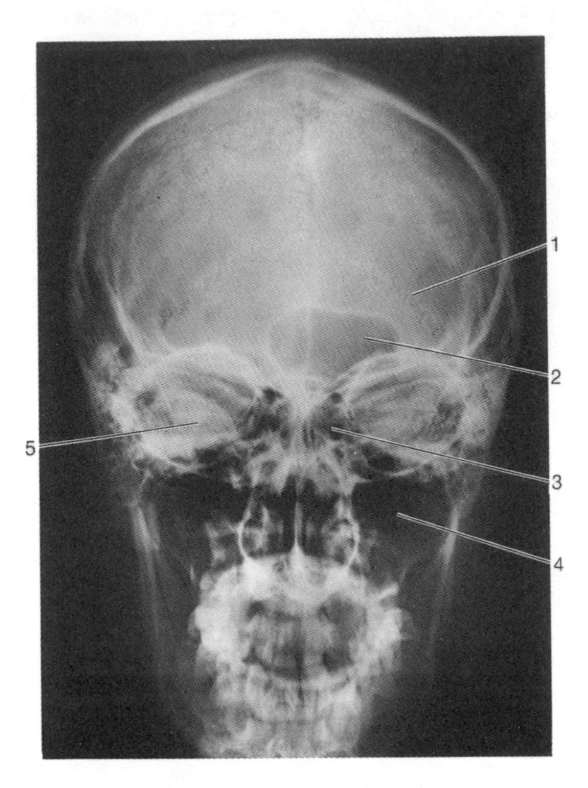

Fig. 4.2 Posteroanterior radiograph of skull. By kind permission of Churchill Livingstone, Publishers. From *Clinical Anatomy in Action*, Volume 2: Pegington.

(a) The serrated line '1' is the coronal suture.
(b) The translucent area '2' is made by the frontal sinus.
(c) Label '3' marks the position of the ethmoidal air cells.
(d) The sinus responsible for the large translucent area '4' drains into the nose beneath the inferior concha.
(e) The opaque shadow '5' obscuring details of the orbit is the greater wing of the sphenoid.

89 Concerning the lateral wall of the nose
(a) the sphenoidal air sinus opens beneath the middle concha.
(b) the nasolacrimal duct opens beneath the inferior concha.
(c) the frontal air sinus opens into the spheno-ethmoidal recess.
(d) the vomer is one of the bones involved in its formation.
(e) the anterior ethmoidal nerve carries sensation to the mucosa of the wall.

88 (a) **False** This is the lambdoid suture.
(b) **True** The frontal sinuses are often asymmetrical in shape.
(c) **True** The ethmoidal air cells lie between the nose and the medial wall of the orbit.
(d) **False** The maxillary air sinus opens beneath the middle concha.
(e) **False** The opaque shadow is made by the petrous temporal bone. Radiographs with the head tilted in the sagittal plane are required if views are to be obtained in which the petrous temporal bones do not obscure the orbits.

89 (a) **False** The air sinus opens in the spheno-ethmoidal recess in the roof of the nose.
(b) **True** The bony canal conducting the duct is made from lacrimal bone, maxilla and the inferior concha.
(c) **False** The sinus opens into the hiatus semilunaris beneath the middle concha.
(d) **False** The vomer is found in the septum of the nose.
(e) **True** Sensation is also carried in the anterior superior alveolar nerve, and in nasal and palatine branches of the sphenopalatine ganglion.

90 Branches from the intrapetrous part of the 7th cranial nerve include
(a) the nerve to stapedius.
(b) the chorda tympani.
(c) the nerve to tensor tympani.
(d) the greater petrosal nerve.
(e) the auriculotemporal nerve.

91 The external jugular vein
(a) receives blood from the retromandibular vein.
(b) pierces the investing layer of deep cervical fascia.
(c) is separated from the surface of the sternocleidomastoid by deep cervical fascia.
(d) contains valves.
(e) drains into the subclavian vein.

92 Bones found on the interior of the skull in the middle cranial fossa include
(a) the frontal bone.
(b) the ethmoid.
(c) the greater wing of the sphenoid.
(d) the petrous part of the temporal bone.
(e) the mastoid process.

93 In the neck the brachial plexus
(a) has its roots (C5, 6, 7, 8 and T1) located in front of scalenus anterior.
(b) forms trunks in the posterior triangle of the neck.
(c) has a suprascapular nerve arising from its upper trunk.
(d) is crossed by the superficial cervical (transverse cervical) artery.
(e) is located in front of the prevertebral fascia.

90 (a) **True** In Bell's palsy (7th cranial nerve palsy) damage to this nerve leads to hyperacusis – pain in the ear on exposure to loud sounds.
(b) **True** Damage to this nerve in Bell's palsy results in loss of taste on the anterior two-thirds of the tongue and loss of secretion in the submandibular gland (on the same side as the lesion).
(c) **False** This muscle is supplied by the mandibular nerve.
(d) **True** Damage to this nerve results in less tear production (same side as the lesion).
(e) **False** This is a branch of the mandibular nerve.

91 (a) **True** The vein is formed by the union of the posterior division of the retromandibular vein and the posterior auricular vein.
(b) **True** The vein pierces the fascia to reach the subclavian vein.
(c) **True** The sternomastoid is invested by deep cervical fascia, the vein runs obliquely downwards on the surface of the fascia.
(d) **True** The vein usually contains two pairs of valves.
(e) **True** The usual termination of the vein is into the subclavian vein.

92 (a) **False** The frontal bone is found in the anterior cranial fossa.
(b) **False** This bone is located in the anterior cranial fossa.
(c) **True** This forms a major part of the fossa and contains several foramina. Between it and the lesser wing is the superior orbital fissure.
(d) **True** The anterior slope of the petrous temporal bone forms part of the fossa floor.
(e) **False** The process does not form part of the floor of the fossa.

93 (a) **False** The roots of the plexus lie in a plane between the scalenus anterior and medius.
(b) **True** Upper (C5, C6), middle (C7), and lower (C8, T1) trunks are located in the posterior triangle of the neck.
(c) **True** The point at which this nerve arises from the trunk is known as Erb's point.
(d) **True** This vessel and the suprascapular artery are branches of the thyrocervical trunk of the subclavian artery.
(e) **False** The prevertebral fascia is continued along the plexus, axillary artery and axillary vein as the fibrous axillary sheath (Gray and Cunningham). Some authorities do not include the vein in the sheath (Last).

94 An inexperienced young doctor attempts to remove a cyst from the side of a patient's neck and accidentally cuts the left hypoglossal nerve. Examination of the patient immediately after the operation reveals
(a) a loss of function of the left mylohyoid muscle.
(b) weakness of the left muscles of mastication.
(c) the tongue deviates to the left when an attempt is made to protrude it.
(d) taste is lost on the anterior two-thirds of the tongue (left side).
(e) atrophy (wasting) of the left half of the tongue muscles.

95 The omohyoid muscle
(a) has an inferior belly arising from the coracoid process of the scapula.
(b) has an intermediate tendon attached to the hyoid bone by a sling of fascia.
(c) is supplied by branches of the ansa cervicalis.
(d) assists in depression of the hyoid.
(e) is found in both anterior and posterior triangles of the neck.

96 Parasympathetic fibres pass from the sphenopalatine (pterygopalatine) ganglion to
(a) the parotid.
(b) the lacrimal gland.
(c) the submandibular gland.
(d) glands in the mucosa of the maxillary air sinus.
(e) taste buds on the posterior third of the tongue.

97 Concerning the laryngeal muscles
(a) The cricothyroid muscle ABducts the vocal cord.
(b) The posterior cricoarytenoid is an ADDuctor of the cords.
(c) The vocalis is supplied by the external laryngeal nerve.
(d) The aryepiglottic muscles form part of the inlet of the larynx.
(e) The lateral cricoarytenoid muscles are ADDuctors of the cords.

94 (a) **False** The mylohyoid is supplied via the mandibular nerve.
(b) **False** These are supplied by the mandibular nerve.
(c) **True** An attempt to protrude the tongue results in deviation to the side of the lesion.
(d) **False** Taste is supplied via fibres which pass along the lingual nerve to the chorda tympani.
(e) **False** Examination immediately after operation would not reveal wasting – this takes a little time to become evident.

95 (a) **False** The belly arises from the superior border of the scapula.
(b) **False** It is the digastric which has a sling of this sort attached to the hyoid bone.
(c) **True** The muscle receives fibres from C1, C2 and C3 via the ansa.
(d) **True** The other depressors are the sternohyoid and thyrohyoid. The sternothyroid depresses the larynx.
(e) **True** The inferior belly is found in the posterior triangle and the superior belly in the anterior triangle of the neck.

96 (a) **False** Parasympathetic fibres reach the parotid from the otic ganglion.
(b) **True** The pathway involves passage along the zygomatic branch of the maxillary nerve, along the lateral wall of the orbit and finally along the lacrimal nerve.
(c) **False** Parasympathetic fibres for this gland synapse in the submandibular ganglion.
(d) **True** Fibres from the ganglion pass to glands in the nasopharynx, nose and maxillary air sinus.
(e) **False** Taste buds do not receive parasympathetic fibres – they receive sensory fibres.

97 (a) **False** The cricothyroid is a tensor of the cords.
(b) **False** This muscle is the only ABDuctor of the cords in the larynx.
(c) **False** The vocalls is supplied by the recurrent paryngeal nerve.
(d) **True** The inlet is formed by the epiglottis and the aryepiglottic folds.
(e) **True** The other ADDuctor is the transverse arytenoid.

98 Use the following coronal CT to answer the question.

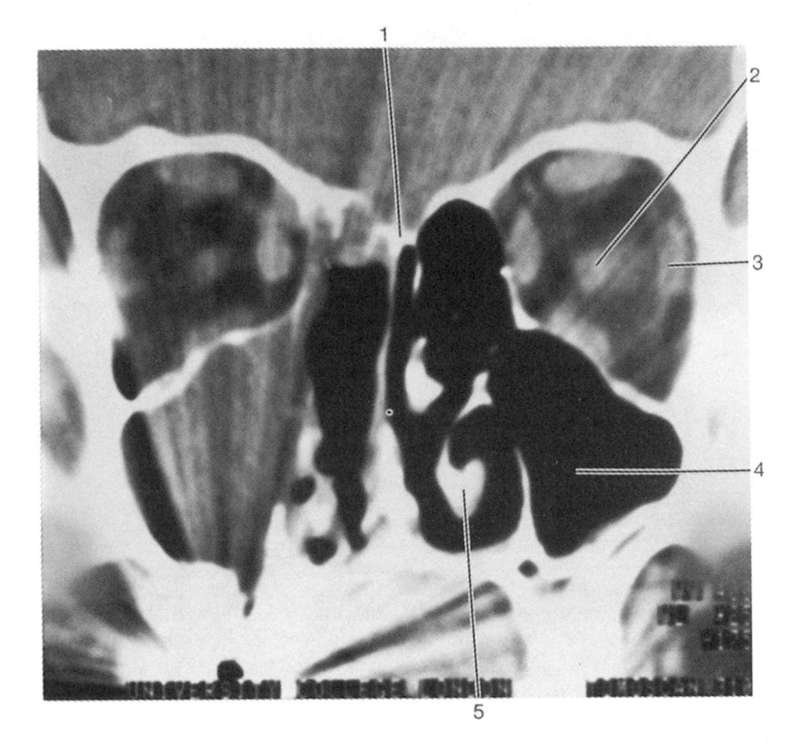

Fig. 4.3 Coronal CT through orbits. By kind permission of Churchill Livingstone, Publishers. From *Clinical Anatomy in Action*, Volume 2: Pegington.

(a) Cranial nerve fibres pass through foramina in bone '1'.
(b) Structure '2' is a cranial nerve.
(c) Structure '3' is supplied by the oculomotor nerve.
(d) Translucent area '4' is a cavity within the maxilla.
(e) Structure '5' is one of the nasal conchae.

99 The left recurrent laryngeal nerve
(a) supplies the left cricothyroid muscle.
(b) contains sensory fibres.
(c) is closely related to the superior thyroid artery.
(d) may be found in the groove between the oesophagus and trachea.
(e) is closely related to the ligamentum arteriosum.

98 (a) **True** The bone is the cribriform plate of the ethmoid and olfactory nerve fibres pass through its foramina.
(b) **True** This is the optic nerve.
(c) **False** The structure is the lateral rectus muscle and this is supplied by the abducent nerve.
(d) **True** This is the maxillary air sinus.
(e) **True** This is one of the nasal conchae, probably the inferior.

99 (a) **False** This is the one intrinsic muscle not supplied by the recurrent laryngeal nerve. It is supplied by the external laryngeal nerve.
(b) **True** These supply sensation to mucosa below the level of the vocal cords.
(c) **False** It is the external laryngeal nerve which is related to this vessel: the recurrent nerve is related to the inferior thyroid artery.
(d) **True** This is the location to look for the nerve on both sides.
(e) **True** The left recurrent laryngeal nerve hooks under the aortic arch immediately behind the attachment of the ligamentum arteriosum.

100 Nerves which pass in the LATERAL WALL of the cavernous sinus include
(a) the mandibular division of V.
(b) the VI nerve.
(c) the ophthalmic division of V.
(d) the anterior ethmoidal nerve.
(e) the III nerve.

101 The trigeminal ganglion
(a) contains synapses.
(b) lies in Meckel's cave.
(c) lies near the apex of the petrous temporal bone.
(d) lies in close relationship to the motor root of the trigeminal nerve.
(e) is found in the posterior cranial fossa.

102 The falx cerebri
(a) is attached to the crista galli.
(b) contains the straight sinus along its attachment to the tentorium cerebelli.
(c) has the superior petrosal sinus in its free margin.
(d) lies in the mid-line between the two cerebellar hemispheres.
(e) has the superior sagittal sinus in its upper margin.

103 The accessory nerve
(a) is a mixed nerve.
(b) has a spinal root for the supply of sternocleidomastoid and trapezius.
(c) has a cranial root which ascends through the foramen magnum.
(d) has fibres which are distributed with pharyngeal and laryngeal branches of the vagus.
(e) leaves the skull through the jugular foramen.

104 The nasociliary nerve
(a) is a branch of the ophthalmic division of V.
(b) enters the orbit through the optic foramen.
(c) supplies skin on the nose.
(d) supplies mucous membrane in the nose.
(e) has branches called the short ciliary nerves.

100 (a) **False** The maxillary division, however, runs in the lower part of the wall of the sinus.
(b) **False** The sixth nerve runs through the meshwork of the sinus with the internal carotid artery – not in the lateral wall.
(c) **True** The ophthalmic division runs in the wall just above the maxillary division.
(d) **False** The nerve only reaches the anterior cranial fossa and its intracranial course is short: it descends through the cribriform plate of the ethmoid to reach the nose.
(e) **True** The oculomotor nerve lies above the trochlear nerve in the lateral wall of the sinus.

101 (a) **False** The trigeminal ganglion is a sensory ganglion and therefore contains cell bodies.
(b) **True** Meckel's, or the trigeminal, cave is a recess of dura.
(c) **True** An impression is usually evident in this position on a dried skull.
(d) **True** The motor root lies inferior to the ganglion.
(e) **False** The ganglion lies in the middle cranial fossa.

102 (a) **True** This bony crest is part of the ethmoid.
(b) **True** It receives the inferior sagittal sinus, the great cerebral vein and some superior cerebellar veins.
(c) **False** The inferior sagittal sinus lies in its free edge.
(d) **False** The falx lies in the longitudinal fissure between the two cerebral hemispheres.
(e) **True** The sinus begins just behind the crista galli and runs backwards in the attached margin of the falx to the occiput. Here it is usually continued as the right transverse sinus.

103 (a) **False** The accessory nerve is a motor nerve.
(b) **True** The spinal root supplies motor fibres to these two muscles.
(c) **False** It is the spinal root which ascends through the foramen magnum.
(d) **True** Fibres of the cranial root are distributed through these vagal branches.
(e) **True** The nerve runs through the middle part of the foramen, the spinal part running in the same dural sheath as the vagus: the cranial part joins the vagus.

104 (a) **True** The other branches are the frontal and lacrimal nerves.
(b) **False** The optic nerve and ophthalmic artery pass through this foramen, the nasociliary nerve passes through the superior orbital fissure.
(c) **True** The terminal part of the nerve is the external nasal branch.
(d) **True** It supplies sensation on part of the nasal septum and lateral wall of the nose.
(e) **False** The short ciliary nerves are branches of the ciliary ganglion. The long ciliary nerves are branches of the nasociliary nerve and contain sympathetic fibres for the dilator pupillae and sensory fibres (including cornea).

105 Use the following diagram of the right orbit to answer the question.

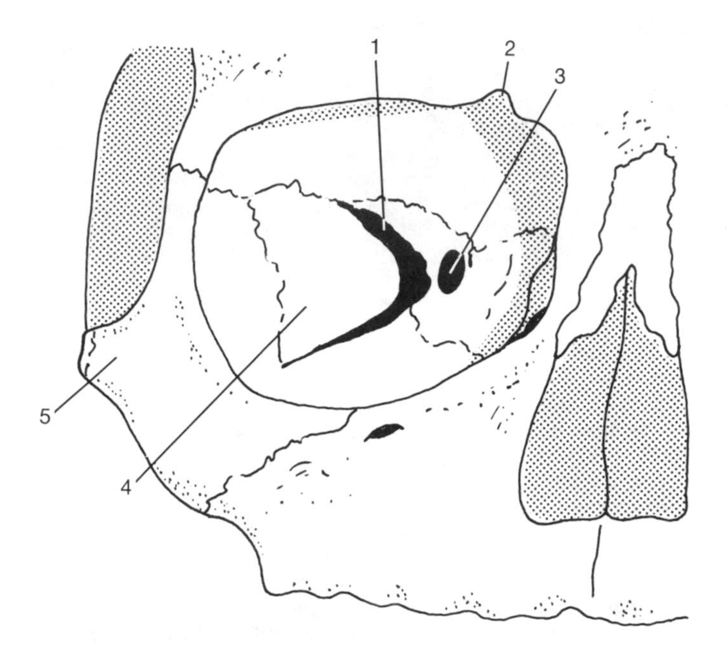

Fig. 4.4 Diagram of the bones of the right orbit.

(a) The 4th cranial nerve passes through the slit marked '1'.
(b) A branch of the frontal nerve travels through the notch marked '2'.
(c) The foramen '3' conducts the ophthalmic artery.
(d) Bone '4' is the ethmoid.
(e) Bone '5' is pierced by the zygomaticofacial nerve.

106 The ciliary ganglion
(a) is a sensory ganglion.
(b) receives parasympathetic fibres from the 3rd nerve.
(c) transmits sympathetic motor fibres to the ciliary muscle.
(d) transmits sympathetic fibres from the internal carotid plexus.
(e) transmits sensory fibres from the eyeball to the nasociliary nerve.

105 (a) **True** The slit is the superior orbital fissure.
(b) **True** The supra-orbital nerve passes through the notch or foramen.
(c) **True** This is the optic foramen.
(d) **False** Bone '4' is the greater wing of the sphenoid.
(e) **True** This is the zygomatic bone and the nerve runs through a bony canal in it.

106 (a) **False** It is classified as a parasympathetic ganglion since these are the only fibres which synapse there.
(b) **True** They come from a small branch of the inferior division of the oculomotor nerve.
(c) **False** The motor fibres to this muscle are parasympathetic.
(d) **True** The fibres pass without synapse through the ganglion to the blood vessels of the eye. Most of the fibres to the dilator pupillae pass via the nasociliary nerve and long ciliary nerves, although a few may pass through the ganglion.
(e) **True** The sensory root of the ganglion joins the nasociliary nerve.

107 The auriculotemporal nerve
- (a) is a branch of the mandibular nerve.
- (b) receives secretomotor fibres from the otic ganglion.
- (c) supplies sensation over the outer surface of the tympanic membrane.
- (d) is located between the neck of the mandible and the sphenomandibular ligament.
- (e) embraces the origin of the middle meningeal artery.

108 Foramina within the sphenoid bone include
- (a) the foramen ovale.
- (b) the internal acoustic meatus.
- (c) the pterygoid canal.
- (d) the foramen spinosum.
- (e) the foramen rotundum.

109 Branches arising from the LEFT vagus nerve IN THE NECK include
- (a) a superior laryngeal branch.
- (b) a recurrent laryngeal branch.
- (c) a meningeal branch.
- (d) cardiac branches.
- (e) a branch to the pharyngeal plexus.

110 The superior cervical sympathetic ganglion
- (a) gives branches which form a plexus around the internal carotid artery.
- (b) sends fibres to the constrictor pupillae.
- (c) gives a cardiac branch.
- (d) may be located on the neck of the first rib.
- (e) communicates with the middle cervical ganglion by means of the ansa subclavia.

107 (a) **True** It is a branch of the posterior trunk of the nerve.
(b) **True** They enter the parotid gland.
(c) **True** The tympanic branch of the vagus also supplies sensation to this surface and carries a few facial nerve fibres.
(d) **True** From here, the nerve passes behind the TMJ.
(e) **True** This pattern is often found.

108 (a) **True** The foramen conducts the mandibular nerve, the accessory meningeal artery and the lesser petrosal nerve.
(b) **False** This foramen is found in the petrous temporal bone.
(c) **True** It conducts the nerve of the pterygoid canal (a mixture of sympathetic fibres with parasympathetic fibres from the greater petrosal nerve).
(d) **True** The foramen transmits the middle meningeal artery.
(e) **True** This foramen transmits the maxillary division of the trigeminal nerve.

109 (a) **True** This divides into internal (sensory) and external (motor) branches.
(b) **False** The left recurrent laryngeal nerve arises from the vagus in the thorax.
(c) **True** This arises from the nerve at the level of the superior ganglion.
(d) **True** Superior and inferior cardiac branches arise in the neck.
(e) **True** This pharyngeal branch transmits motor fibres to the plexus (cranial accessory fibres).

110 (a) **True** The internal carotid nerve breaks up to form the internal carotid plexus, and transmits sympathetic fibres to structures in the head.
(b) **False** The sympathetic fibres supply the dilator pupillae.
(c) **True** Both right and left ganglia give a cardiac branch.
(d) **False** This is the location for the cervicothoracic or stellate ganglion.
(e) **False** The ansa subclavia connects the middle and inferior ganglia.

The Upper Limb

111 Use the radiograph of the elbow to answer the question.

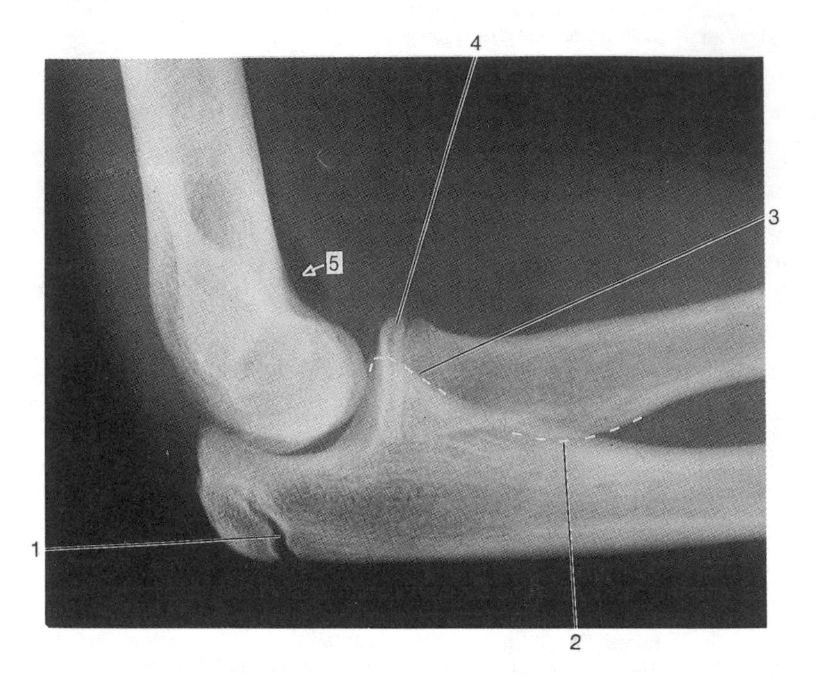

Fig. 5.1 Radiograph of elbow. By kind permission of Churchill Livingstone, Publishers. From *Clinical Anatomy in Action*, Volume 1: Pegington.

(a) the radiolucent line '1' is an epiphyseal line.
(b) the section of radius outlined with dots at '2' gives insertion to a tendon.
(c) the section of the ulna outlined with dots at '3' is the coronoid process.
(d) the epiphysis labelled '4' first appears during the first year of life.
(e) the radiolucent soft tissue area labelled '5' is the brachial artery.

112 If the median nerve is completely severed 2.5 cm proximal to the flexor retinaculum
(a) a 'claw hand' deformity results.
(b) the abductor pollicis brevis will be paralysed.
(c) all the interossei will be paralysed.
(d) there will be loss of sensation over the palmar surfaces and nail beds of the radial three and a half digits.
(e) there will be some loss of sensation in the palm.

111 (a) **True** The epiphysis for the olecranon is first seen on radiographs at about the age of 9–11 years.
(b) **True** The bony process is the radial tuberosity and it gives insertion to the tendon of biceps.
(c) **True** The brachialis is inserted into the process.
(d) **False** The epiphysis at the lower end of the radius usually appears at the end of the first year but the head starts to ossify at the fourth year in females and fifth year in males.
(e) **False** The brachial artery does not normally cast a shadow. If a radiolucent area is seen in front of or behind the lower humerus in this region it is produced by one of the intra-capsular fat pads. When there is an excess of synovial fluid after injury the extra-synovial pads are pushed away from the bone.

112 (a) **False** A 'claw hand' is a feature of ulnar nerve damage. In median nerve injury the deformity is an 'ape-like' hand, because of inability to oppose the thumb.
(b) **True** This is best tested by laying the injured hand with the dorsum on a table, and the subject is asked to abduct the thumb and touch a pen held just above the hand.
(c) **False** The interossei are supplied by the ulnar nerve.
(d) **True** Cutaneous branches of the median nerve supply these areas.
(e) **True** The palmar cutaneous branch arises just proximal to the flexor retinaculum and crosses its surface to reach the palm.

113 Use the following diagram of the front of the left humerus to answer the question.

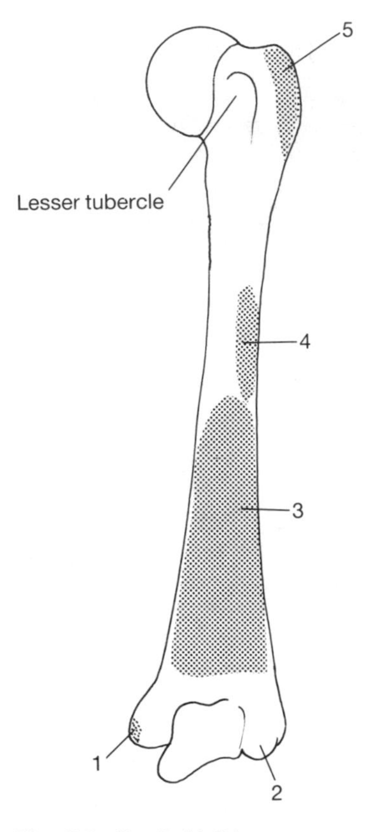

Fig. 5.2 Front of left humerus.

(a) Brachioradialis attaches to bony prominence '1'.
(b) The articular surface '2' is the trochlea.
(c) Brachialis attaches at '3'.
(d) Pectoralis major attaches at '4'.
(e) Subscapularis attaches to the bony prominence '5'.

114 The lumbrical muscle of the index finger
(a) arises from a metacarpal.
(b) inserts into a dorsal expansion.
(c) can extend the index finger at the metacarpophalangeal joint.
(d) is supplied by the median nerve.
(e) passes along the ulnar (medial) side of the metacarpophalangeal joint.

113 (a) **False** This is the medial epicondyle which gives origin to flexor and pronator muscles. Brachioradialis arises just above the LATERAL epicondyle.
(b) **False** This is the capitulum.
(c) **True** The brachialis arises from much of the front of the humerus as high as the insertion of deltoid.
(d) **False** This is the position for the deltoid insertion.
(e) **False** This is the greater tubercle: subscapularis inserts into the lesser tubercle.

114 (a) **False** Each lumbrical arises from its corresponding flexor digitorum profundus tendon.
(b) **True** The delicate tendon inserts along the radial edge of the expansion.
(c) **False** The lumbrical acts as a weak flexor at the MCP joint.
(d) **True** The radial two lumbricals are usually supplied by the median nerve and the ulnar two by the ulnar nerve.
(e) **False** The lumbricals pass along the radial sides of the MCP joints.

115 Use the following diagram of the front of the right radius and ulna to answer the question.

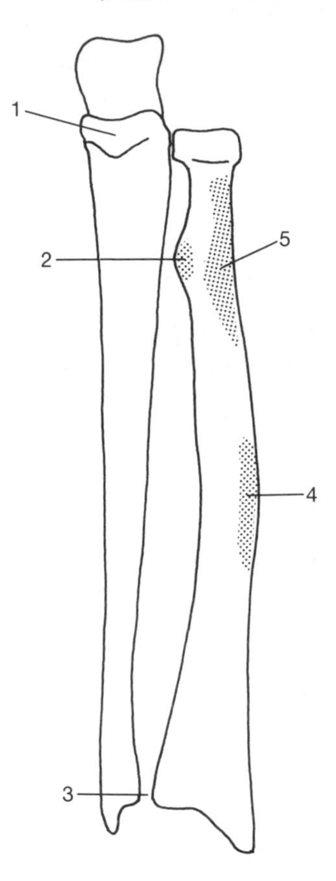

Fig. 5.3 Front of right radius and ulna.

(a) The biceps attaches to the bony process '1'.
(b) A bursa is associated with the attachment of the tendon to '2'.
(c) The joint '3' is of the secondary cartilaginous type.
(d) Area '4' gives attachment to pronator teres.
(e) The muscle which attaches to '5' is supplied by the median nerve.

116 When the deep branch of the ulnar nerve has been destroyed by compression from a ganglion, the patient presents with
(a) paralysis of the abductor pollicis brevis.
(b) an ape-like deformity of the hand.
(c) paralysis of the adductor pollicis.
(d) loss of sensation over the palmar surface of the little finger.
(e) loss of sensation on the dorsum of the hand.

115 (a) **False** The brachialis is attached to the coronoid process of the ulna.
(b) **True** A bursa is found close to the insertion of the biceps tendon.
(c) **False** The inferior radio-ulnar joint is synovial.
(d) **True** Pronator teres is inserted into the middle of the lateral surface of the radius.
(e) **False** Supinator is attached to the radius here and is supplied by the radial nerve.

116 (a) **False** This muscle is supplied by the median nerve.
(b) **False** A claw hand results (*see also* answer to Q.112)
(c) **True** The deep branch of the ulnar nerve supplies the hypothenar muscles, the interossei, 3rd and 4th lumbricals and the adductor pollicis.
(d) **False** This is supplied by branches of the superficial branch of the ulnar nerve.
(e) **False** The dorsal cutaneous branch of the ulnar nerve arises from the main trunk in the distal forearm.

117 When the radial nerve is compressed in the axilla the subject presents with
- (a) inability to extend the index finger.
- (b) paralysis of extensor digiti minimi.
- (c) inability to supinate the forearm when the elbow is flexed to a right angle.
- (d) paralysis of brachioradialis.
- (e) wrist drop.

118 Structures attached to the coracoid process of the scapula include
- (a) the long head of biceps.
- (b) a ligament of the shoulder joint.
- (c) pectoralis major.
- (d) a ligament which protects the acromioclavicular joint from dislocation.
- (e) the omohyoid.

119 Concerning the muscles of the rotator (tendinous) cuff:
- (a) infraspinatus acts as a medial rotator at the shoulder.
- (b) the subscapularis is intimately related to the subacromial bursa.
- (c) the supraspinatus is supplied by a branch of the posterior cord of the brachial plexus.
- (d) the supraspinatus is separated from the coraco-acromial ligament by the subacromial bursa.
- (e) teres minor is supplied by the axillary nerve.

120 Injury to the axillary nerve
- (a) sometimes occurs during anterior dislocation of the shoulder.
- (b) occasionally occurs when the humerus is fractured across its surgical neck.
- (c) results in loss of sensation in the skin over the lower deltoid.
- (d) eventually results in an abnormal contour of the shoulder.
- (e) results in an inability to abduct the arm at the shoulder above the head.

121 The biceps brachii
- (a) arises by a long head from the superior border of the scapula.
- (b) is a supinator.
- (c) is a flexor of the arm at the elbow.
- (d) is supplied in part by the radial nerve.
- (e) inserts into the ulna by means of a strong tendon.

117 (a) **True** The radial nerve, through its posterior interosseous branch, supplies both extensor digitorum and extensor indicis.
(b) **True** This muscle is also supplied by the posterior interosseous nerve.
(c) **False** Supination of the forearm with the elbow in the flexed position is a function of biceps brachii, and this is supplied by the musculocutaneous nerve.
(d) **True** The muscle is supplied by the radial nerve.
(e) **True** The extensor musculature of the wrist is supplied by the radial nerve.

118 (a) **False** The short head is attached to the process.
(b) **True** The coracohumeral ligament gains attachment to the lateral border of the process.
(c) **False** The pectoralis minor is attached to the process.
(d) **True** Both conoid and trapezoid parts of the coracoclavicular ligament are attached to the process.
(e) **False** The posterior belly of omohyoid is attached to the upper border of the scapula.

119 (a) **False** This muscle is a lateral rotator at the shoulder.
(b) **False** The subacromial bursa is related to the supraspinatus: the subscapular bursa lies deep to the subscapularis.
(c) **False** The nerve to supraspinatus, the suprascapular nerve, is a branch of the upper trunk of the brachial plexus.
(d) **True** The bursa lies deep to the ligament and the deltoid and is closely related to the supraspinatus. Calcification in the latter muscle may lead to bursitis.
(e) **True** The axillary nerve supplies a muscular twig to this muscle.

120 (a) **True** The nerve runs just below the shoulder joint capsule.
(b) **True** The nerve is closely related to the surgical neck of the humerus and may be damaged during fracture at this site.
(c) **True** The upper lateral cutaneous nerve, a branch of the axillary nerve, supplies skin over the lower deltoid and upper part of the long head of triceps.
(d) **True** This is due to wasting of the deltoid. The acromion and greater tubercle are both easily visible, the latter being the most lateral of the two in position.
(e) **True** Paralysis of the deltoid results in an inability to abduct the arm. Supraspinatus cannot fulfil this function.

121 (a) **False** The long head arises from the supraglenoid tubercle.
(b) **True** With the elbow in the flexed position the biceps is a powerful supinator: the supinator performs this function when the forearm is straightened.
(c) **True** The muscle, with brachialis and brachioradialis are flexors of the arm at the elbow.
(d) **False** The muscle is supplied by the musculocutaneous nerve.
(e) **False** The strong tendon inserts into the radial tuberosity of the radius: the bicipital aponeurosis fuses with deep fascia.

122 Use the following radiograph of the wrist to answer the question.

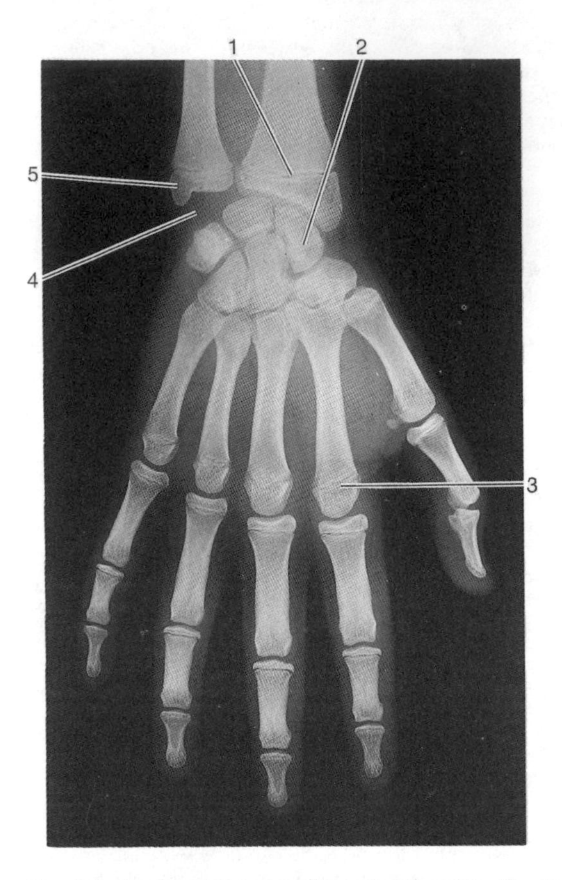

Fig 5.4 Posteroanterior radiograph of wrist. By kind permission of Churchill Livingstone, Publishers. From *Clinical Anatomy in Action*, Volume 1: Pegington.

(a) The radiolucent line '1' is a fracture of the radius.
(b) The radial artery is closely related to bone '2'.
(c) It is quite normal to have an epiphyseal line (growth disc) '3' at the distal end of the index metacarpal.
(d) The 'space' labelled '4' between the carpus and the ulna is normal.
(e) Bone '5' is the triquetrum.

123 The subacromial bursa
(a) is continuous with the synovial membrane of the shoulder joint.
(b) lies deep to the deltoid.
(c) is intimately related to the supraspinatus.
(d) protrudes through a hole in the shoulder joint capsule.
(e) is usually the smallest bursa in the upper limb.

122 (a) **False** This is an epiphyseal line.
(b) **True** The artery runs deep in the anatomical snuffbox, close to the scaphoid (bone '2').
(c) **True** But note that the epiphysis is normally at the BASE of the thumb metacarpal.
(d) **True** The 'space' is occupied by the articular disc.
(e) **False** This is the epiphysis of the lower end of the ulna.

123 (a) **False** The subscapular bursa is continuous with the joint synovium.
(b) **True** The bursa is large and lies deep to both the coraco-acromial arch and the deltoid.
(c) **True** The bursa wall is fused with the tendon of supraspinatus and the rotator cuff.
(d) **False** The subscapular bursa protrudes through a hole which lies between the superior and middle glenohumeral ligaments.
(e) **False** The bursa is by far the largest in the limb, if not in the body.

124 The scaphoid bone of the wrist
(a) gives attachment to the flexor carpi ulnaris.
(b) articulates with only one bone of the carpus.
(c) gives attachment to the flexor retinaculum.
(d) gives attachment to flexor carpi radialis.
(e) receives nutrient vessels through its waist.

125 Concerning the 'anatomical snuffbox':
(a) it is bounded laterally by the tendon of extensor pollicis longus.
(b) the basilic vein lies in the subcutaneous tissues covering the box.
(c) the tubercle of the scaphoid may be palpated in the box.
(d) branches of the radial nerve lie in the superficial tissues covering the box.
(e) pulsations of the radial artery may be palpated in the box.

126 The carpal tunnel transmits
(a) the median nerve.
(b) the ulnar nerve.
(c) the tendons of flexor digitorum superficialis.
(d) the radial artery.
(e) the palmar cutaneous branches of median and ulnar nerves.

127 Flexor digitorum superficialis
(a) arises in part from the lateral epicondyle of the humerus.
(b) arises from both radius and ulna.
(c) flexes the terminal phalanges of the fingers.
(d) splits into four tendons.
(e) is supplied on its ulnar side by the ulnar nerve.

124 (a) **False** The tendon of this muscle is attached to the fifth metacarpal and the hook of the hamate.
(b) **False** The scaphoid articulates with the trapezium, trapezoid, capitate, lunate and radius.
(c) **True** The radial side of the retinaculum gains attachment to the scaphoid and the trapezium.
(d) **False** The tendon of this muscle inserts into the second metacarpal.
(e) **True** This part of the bone is non-articular and nutrient vessels are able to enter.

125 (a) **True** This tendon takes an oblique course from the radial tubercle (Lister's tubercle) to the terminal phalanx of the thumb.
(b) **False** The cephalic vein lies over the snuffbox.
(c) **False** The waist of the bone is palpated in the snuffbox: the tubercle lies in front of the snuffbox.
(d) **True** These are cutaneous branches of the superficial branch of the nerve.
(e) **True** The artery crosses the floor of the snuffbox.

126 (a) **True** Pressure on the nerve as it runs through the tunnel results in carpal tunnel syndrome.
(b) **False** The ulnar nerve travels superficial to the retinaculum, although it is often protected by a separate fibrous sheet.
(c) **True** The tendons of both flexor digitorum superficialis and flexor digitorum profundus pass through the tunnel.
(d) **False** The radial artery lies in front of the lower radius and then passes through the anatomical snuffbox.
(e) **False** Both these nerves pass superficial to the retinaculum.

127 (a) **False** The humero-ulnar head arises from the medial epicondyle and ulna.
(b) **True** The muscle has both humero-ulnar and radial heads.
(c) **False** The tendons of superficialis insert into middle phalanges and therefore cannot flex at the terminal interphalangeal joints.
(d) **True** Each tendon also splits near its insertion to allow the profundus tendon to continue to its insertion.
(e) **False** The muscle is supplied by the median nerve.

128 Concerning the brachial plexus:
   (a) a group of lymph nodes is associated with a branch of the posterior cord.
   (b) a branch of the lateral cord supplies biceps brachii.
   (c) the point of origin of the suprascapular nerve from the plexus lies in the posterior triangle of the neck.
   (d) damage to the nerve to latissimus dorsi results in the condition of 'winging' of the scapula.
   (e) the median nerve is formed from both medial and lateral cords.

129 After complete section of the ulnar nerve behind the medial epicondyle, examination of the affected side will reveal that
   (a) the subject will be unable to flex the distal interphalangeal joint of the index finger.
   (b) the subject will be unable to use adductor pollicis on the affected side.
   (c) the subject's hand will deviate to the ulnar side when an attempt is made to flex the wrist.
   (d) there is loss of sweating over the ulnar half of the palm when the hand is placed in a hot environment.
   (e) there is wrist drop.

130 The female breast
   (a) has an axillary tail.
   (b) receives blood from both axillary and internal thoracic arteries.
   (c) has a nipple located at the level of the 3rd intercostal space.
   (d) has lymph drainage which reaches the apical axillary nodes.
   (e) has lactiferous ducts which open at a single orifice on the nipple.

128 (a) **True** This is the subscapular group.
(b) **True** The musculocutaneous nerve arises from the lateral cord.
(c) **True** The point is known as Erb's point. Damage to the upper trunk during delivery at birth may result in Erb's palsy.
(d) **False** Damage to the nerve to serratus anterior results in winging of the scapula.
(e) **True** The medial cord contribution usually crosses the axillary artery to join the contribution from the lateral cord so that the median nerve lies lateral to the artery.

129 (a) **False** The radial half of flexor digitorum profundus is supplied by the anterior interosseous nerve.
(b) **True** Adductor pollicis is supplied by the deep branch of the ulnar nerve.
(c) **False** The deviation will be towards the radial side because of paralysis of flexor carpi ulnaris.
(d) **True** Sympathetic fibres are carried in the palmar cutaneous branch to the sweat glands of the ulnar half of the palm.
(e) **False** Wrist drop is caused by radial nerve palsy: there will be claw hand. The claw hand of this type of injury is not as marked as the claw of an ulnar nerve injury at the wrist (ulnar nerve paradox). The reason is that the profundus tendons to the clawed ring and little fingers are paralysed in an injury above the elbow – these fingers therefore straighten.

130 (a) **True** This extension must be carefully examined during a physical examination of the female breast.
(b) **True** The superior thoracic and lateral thoracic branches of the axillary artery and perforating branches of the internal thoracic artery (usually 2nd, 3rd and 4th) supply the female breast.
(c) **False** The level of the nipple varies, but is usually at the 4th intercostal space in the nulliparous female.
(d) **True** Lymph from the axillary nodes reaches the apical nodes: these lie close to the coracoid process at the apex of the axilla.
(e) **False** About 15–20 ducts open separately onto the surface of the nipple.

Lower Limb

131 The gluteus maximus
- (a) arises from the ischium.
- (b) inserts into the iliotibial tract.
- (c) inserts into the femur.
- (d) is supplied by the inferior gluteal nerve.
- (e) is a medial rotator of the hip joint.

132 The sciatic nerve
- (a) has a root value of L1, L2 and L3.
- (b) emerges from the pelvis through the lesser sciatic foramen.
- (c) is closely related to the posterior femoral cutaneous nerve.
- (d) divides into tibial and common peroneal branches.
- (e) supplies the muscles of the flexor compartment of the thigh.

133 The femoral sheath
- (a) contains the femoral nerve.
- (b) is a continuation of fascia iliaca and fascia transversalis into the thigh.
- (c) contains the femoral vein in its most medial compartment.
- (d) contains lymphatics.
- (e) contains one of the deep inguinal lymph nodes.

134 The adductor canal
- (a) is situated in the flexor compartment of the thigh.
- (b) is bounded partly by the vastus medialis.
- (c) is roofed by a sheet of fascia and the sartorius muscle.
- (d) transmits the femoral artery and vein.
- (e) transmits the great saphenous vein.

135 The medial meniscus of the knee joint
- (a) is C-shaped.
- (b) has anterior and posterior horns.
- (c) is more mobile than the lateral meniscus.
- (d) gives insertion to some fibres of popliteus.
- (e) is injured more frequently than the lateral meniscus.

131 (a) **False** The muscle arises from the posterior gluteal line and the bone behind and above this line, from the sacrum and coccyx, the sacrotuberous ligament and aponeurosis of erector spinae.
(b) **True** The large superficial part of the muscle is attached to the tract.
(c) **True** The deeper fibres of the muscle are attached to the gluteal tuberosity of the femur.
(d) **True** The gluteus medius and minimus and tensor fasciae latae are supplied by the superior gluteal nerve.
(e) **False** Apart from its powerful action of extending a flexed thigh it can also act as a *LATERAL* rotator.

132 (a) **False** The sciatic nerve is derived from L4, 5, S1, 2 and 3.
(b) **False** The nerve runs through the greater sciatic foramen below the piriformis muscle.
(c) **True** The posterior cutaneous nerve travels on the medial side of the sciatic nerve below the piriformis.
(d) **True** The point of division is usually at the level of the lower third of the thigh, although it may divide at a higher level, and sometimes in the pelvis.
(e) **True** The muscles are the semitendinosus, semimembranosus and biceps femoris.

133 (a) **False** The nerve lies on the lateral side of the sheath, not within it.
(b) **True** These fascial layers are continued over the proximal parts of the femoral artery and vein as the femoral sheath.
(c) **False** The most medial compartment contains fat, lymphatics and the lymph node of Cloquet: it is called the femoral canal. The femoral vein lies in the intermediate compartment of the sheath.
(d) **False** Some lymphatics from the lower limb accompany the femoral vessels and others run through the femoral canal.
(e) **True** The lymph node of Cloquet lies in the femoral canal. It drains lymph from the glans penis (male) or glans clitoridis (female).

134 (a) **False** The canal lies between the quadriceps and the adductor group of muscles in the lower third of the front of the thigh (extensor compartment).
(b) **True** Vastus medialis forms the lateral wall of the canal.
(c) **True** There is a strong sheet of fascia deep to sartorius in the roof of the canal.
(d) **True** Both artery and vein run through the hiatus in the adductor magnus at the lower end of the canal.
(e) **False** The great saphenous vein runs in the subcutaneous tissues.

135 (a) **True** The lateral meniscus forms four-fifths of a complete ring.
(b) **True** The menisci are attached to the tibia by their horns.
(c) **False** The medial meniscus is more firmly fixed than the lateral because of the attachment of the deep fibres of the tibial collateral ligament ot its periphery.
(d) **False** Fibres of popliteus are often attached to the LATERAL meniscus.
(e) **True** The medial meniscus is injured far more frequently than the lateral, partly because it is more fixed and partly because of its shape.

136 Tibialis anterior
- (a) has a tendon which passes deep to the superior extensor retinaculum.
- (b) inserts into the navicular bone.
- (c) extends the foot at the ankle joint.
- (d) assists in maintaining the longitudinal arch of the foot.
- (e) inverts the foot at the subtalar joint.

137 The gastrocnemius
- (a) forms the lower boundary of the popliteal fossa.
- (b) inserts by way of the tendo calcaneus.
- (c) plantarflexes the foot at the ankle joint.
- (d) is surrounded by a synovial sheath near its insertion.
- (e) is supplied by the tibial nerve.

138 During the final few degrees of extension the knee joint 'screws home' or 'locks'. This physiological locking movement . . .
- (a) involves a medial rotation of the femur on the tibia.
- (b) is produced by contraction of the popliteus muscle.
- (c) is accompanied by tension in many of the ligaments of the joint.
- (d) results in the knee adopting the 'close-packed' position.
- (e) is due in part to the medial femoral condyle having a shorter anteroposterior length than the lateral.

139 Bursae are found in and around the knee
- (a) in the subcutaneous tissue in front of the patella.
- (b) closely related to the semimembranosus muscle.
- (c) between the ligamentum patellae and the tibia.
- (d) between the cruciate ligaments.
- (e) deep to the quadriceps.

140 The anterior cruciate ligament of the knee
- (a) attaches to the anterior part of the intercondylar fossa (notch) of the femur.
- (b) is completely surrounded by synovial membrane.
- (c) is associated with meniscofemoral ligaments of Humphry and Wrisberg.
- (d) is stronger and has a greater width than the posterior cruciate ligament.
- (e) plays a part in limiting rotation in the knee.

136 (a) **True** The tendon passes deep to the superior and usually through the inferior extensor retinaculum.
(b) **False** It is tibialis posterior which inserts into this bone, tibialis anterior inserts into the medial cuneiform and the base of the first metatarsal.
(c) **True** The muscle extends or dorsiflexes the foot at the ankle joint.
(d) **True** It contracts powerfully to increase the arch in the take-off phase of walking.
(e) **True** This muscle and the tibialis posterior are powerful invertors of the foot.

137 (a) **True** The two heads form the lower boundaries of the fossa: the upper boundaries are formed by the biceps laterally and the semimembranosus and semitendinosus medially.
(b) **True** The tendo calcaneus (Achilles' tendon) is the common tendon of gastrocnemius, soleus and plantaris.
(c) **True** The gastrocnemius is a powerful plantarflexor of the foot but can also act as a flexor of the knee.
(d) **False** The tendo calcaneus is not surrounded by a synovial sheath.
(e) **False** The muscle is supplied by the tibial nerve.

138 (a) **True** This rotation is a function of the shape of the articular surfaces of the femur (*see* answer (e)).
(b) **False** The popliteus unlocks the knee.
(c) **True** Parts of both cruciate ligaments, collateral ligaments and the oblique posterior ligament are all taut.
(d) **True** In the 'close-packed' position there is maximum congruity between the articular surfaces.
(e) **False** The anteroposterior extent of the medial femoral surface is greater than that of the lateral surface. Thus, when the lateral surface is used up just short of full extension the medial surface continues to slide. This latter movement produces medial rotation of the femur (*see* (a)).

139 (a) **True** Subcutaneous prepatellar and infrapatellar bursae are usually present.
(b) **True** This bursa may become enlarged in certain diseases of the knee.
(c) **True** This is the deep infrapatellar bursa.
(d) **True** A bursa of synovium passes between the ligaments from their lateral aspect.
(e) **True** The suprapatellar bursa is continuous with the synovium of the joint: it extends upwards to 4 finger-breadths above the upper border of the patella.

140 (a) **False** The ligament is attached to the medial surface of the lateral condyle of the femur.
(b) **False** Synovial membrane only covers the sides and front of the cruciate ligaments.
(c) **False** The meniscofemoral ligaments run from the posterior part of the lateral meniscus to the medial femoral condyle: the course takes them in front of, and behind, the posterior cruciate ligament.
(d) **False** It is the posterior ligament which is stronger and wider than the anterior.
(e) **True** Both cruciate ligaments play a part in limiting medial rotation of the tibia (lateral rotation of the femur).

141 Use the following radiograph of the ankle to answer the question

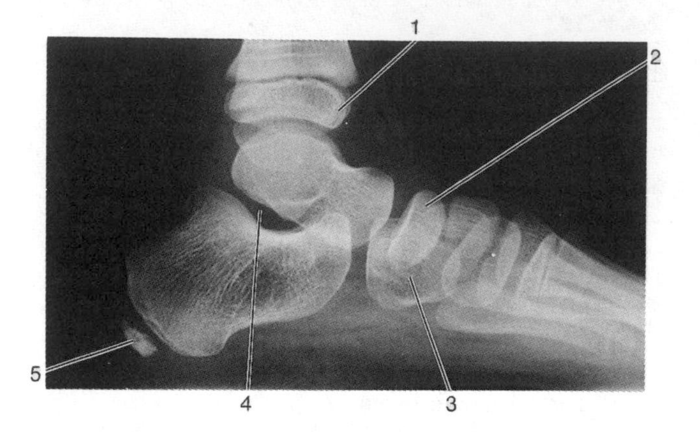

Fig. 6.1 Lateral radiograph of ankle. By kind permission of Churchill Livingstone, Publishers. From *Clinical Anatomy in Action*, Volume 1: Pegington.

(a) Label '1' indicates one of the tarsal bones.
(b) Bone '2' would be visible on a radiograph of a neonatal foot.
(c) Bone '3' is one of the cuneiforms.
(d) Joint space '4' indicates the position of a fibrous joint.
(e) Bone '5' is a sesamoid.

142 The subtalar joint
(a) has articular surfaces covered with fibrocartilage.
(b) is found between the cuboid and calcaneus.
(c) allows movements of inversion and eversion.
(d) is supported on its lateral side by the calcaneofibular ligament.
(e) is visible on a lateral radiograph of an ankle.

143 The great saphenous vein
(a) is located on the front of the lateral malleolus.
(b) is connected to deep veins in the calf.
(c) possesses bicuspid valves.
(d) drains into the femoral vein.
(e) has a tributary called the posterior arch vein.

141 (a) **False** This is the lower epiphysis of the tibia.

(b) **False** Bone '2' is the navicular and this starts its ossification in about the third year of life. The calcaneus and talus do start to ossify before birth and sometimes the cuboid has also started to form by birth.

(c) **False** This is the cuboid.

(d) **False** Label '4' is the subtalar joint space – a synovial joint.

(e) **False** Although the calcaneus starts ossifying before birth a second scale-like epiphysis appears at the back of the bone at 6–8 years of age.

142 (a) **False** The articular surfaces of the subtalar joint are covered with hyaline cartilage.

(b) **False** The subtalar joint lies between the talus and calcaneus.

(c) **True** The movements of inversion and eversion take place at the subtalar and talonavicular joints.

(d) **True** This ligament extends across the ankle and subtalar joints, unlike the other two bands of the lateral ligament – the anterior and posterior talofibular ligaments. These latter ligaments span the ankle joint only.

(e) **True** The convex articular surface of the calcaneus is found on the upper surface of the calcaneus.

143 (a) **False** The vein may be located on the front of the MEDIAL malleolus.

(b) **True** The 'perforating' veins are important clinically: incompetence results in varicosity of the great saphenous vein.

(c) **True** The vein contains some 10–20 valves; most are located below the knee.

(d) **True** The great saphenous vein passes through the saphenous opening to reach the femoral vein. The junction of the two veins is guarded by a valve.

(e) **True** The posterior arch vein is a tributary which runs upwards from the medial malleolus to drain into the great saphenous vein below the knee. Its clinical importance is that it has several perforators which join the venae comitantes of the posterior tibial artery (Dodd and Cockett, 1956).

144 The following diagram represents the medial aspect of the upper tibia and fibula. Use it to answer the following items . . .

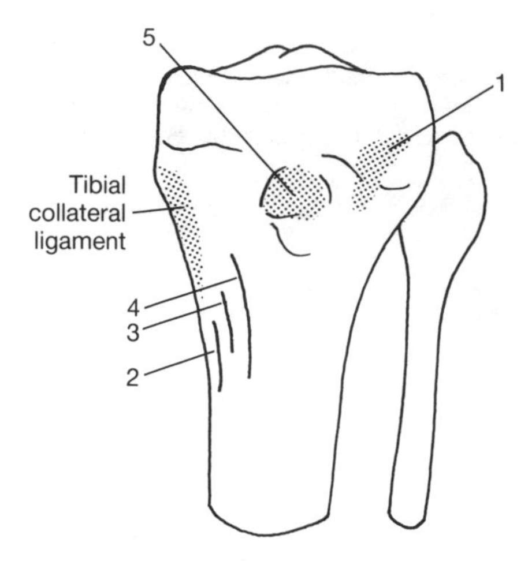

Fig. 6.2 Medial aspect of the upper end of the tibia.

(a) The fibular collateral ligament attaches to the bone at '1'.
(b) The semitendinosus attaches to the bone at '2'.
(c) The iliotibial tract attaches to the bone at '3'.
(d) The sartorius attaches to the bone at '4'.
(e) The ligamentum patellae is attached to the bone at '5'.

145 The lateral ligament of the ankle joint
(a) is sometimes known as the deltoid ligament.
(b) is the ligament injured in the common 'sprained' ankle.
(c) gains attachment to the talus.
(d) gains attachment to the sustentaculum tali.
(e) prevents excessive inversion at the subtalar joint.

144 (a) **False** This maps the attachment of the iliotibial tract.

(b) **True** The semitendinosus tendon is the posterior of a group of the three insertions '2', '3' and '4'.

(c) **False** This is the insertion of gracilis.

(d) **True** The student often remembers these three by the mnemonic 'Say Grace Before Tea' (sartorius, gracilis, a bursa and semitendinosus). The three are sometimes referred to as the pes anserinus.

(e) **True** The ligamentum patellae is attached to the upper smooth part of the tibial tuberosity.

145 (a) **False** It is the medial ligament which is usually referred to as the deltoid ligament.

(b) **True** Injury usually involves only the anterior talofibular band, but the calcaneofibular band may be damaged in a severe sprain.

(c) **True** Both anterior and posterior talofibular bands are attached to the talus.

(d) **False** The sustentaculum, a ridge on the calcaneus, lies on the medial side of the ankle.

(e) **True** Indeed, excessive inversion often results in a 'sprained' ankle. The peroneal muscles are also active in preventing inversion injuries.

146 The tibial collateral ligament of the knee
- (a) is a cord-like fibrous structure.
- (b) is attached to the tibia.
- (c) is attached to the medial meniscus.
- (d) is pierced by the tendon of popliteus.
- (e) is closely related to the common peroneal nerve.

147 The sciatic nerve
- (a) is a branch of the lumbosacral plexus.
- (b) leaves the pelvis through the lesser sciatic foramen.
- (c) enters the gluteal region between obturator internus and quadratus femoris.
- (d) has a surface marking midway between the greater trochanter and the iliac crest.
- (e) lies deep to gluteus maximus.

148 Structures passing deep to the flexor retinaculum of the ankle include
- (a) the deep peroneal nerve.
- (b) the posterior tibial artery.
- (c) the tendon of tibialis anterior.
- (d) the tendon of flexor hallucis longus.
- (e) the great saphenous vein.

146 (a) **False** The tibial collateral ligament is a broad flat band: the fibular collateral ligament is a rounded cord.

(b) **True** The tibial attachment is to the medial condyle and the shaft of the tibia. This latter attachment extends well down the shaft, the ligament being crossed by the tendons of sartorius, gracilis and semitendinosus (*see* Q.141).

(c) **True** Some of the deep fibres gain attachment to the periphery of the medial meniscus.

(d) **False** The tendon of popliteus pierces the capsule of the knee joint, not the collateral ligament.

(e) **False** The common peroneal nerve travels to the LATERAL side of the popliteal fossa. The tibial collateral ligament is closely related to the medial inferior genicular vessels and nerve.

147 (a) **True** The sciatic nerve has a root value of L4, L5, S1, S2, S3 and S4.

(b) **False** The nerve leaves through the greater sciatic foramen.

(c) **False** The nerve leaves at the lower border of piriformis.

(d) **False** The surface marking is half-way between the greater trochanter and the ischial tuberosity.

(e) **True** The nerve is situated deep to the gluteus maximus in its upper course.

148 (a) **False** The deep peroneal nerve travels at the front of the leg and dorsum of the foot, and is not related to the flexor retinaculum.

(b) **True** The posterior tibial artery is accompanied by two veins and the tibial nerve.

(c) **False** The tendon of tibialis POSTERIOR passes deep to the flexor retinaculum.

(d) **True** The structures which pass deep to the retinaculum from medial to lateral are: tibialis posterior, flexor digitorum longus, posterior tibial vessels, tibial nerve, and flexor hallucis longus. The student can remember these with the mnemonic – Tom, Dick and Harry.

(e) **False** The vein runs on the surface of the medial malleolus.

149 The following diagram represents the front of the femur. Use it to answer the question

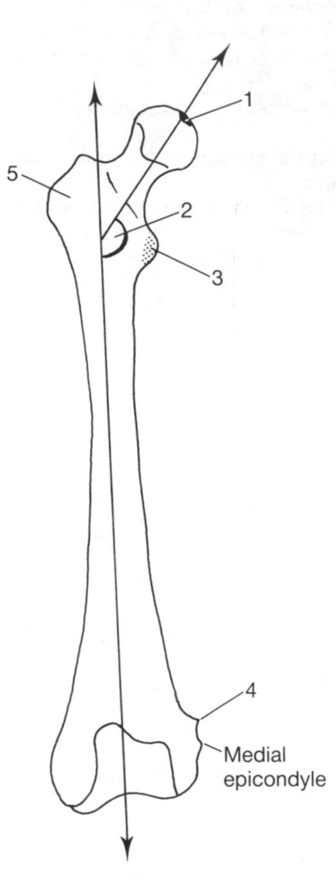

Fig. 6.3 Diagram of the front of the femur.

(a) The hole '1' transmits a blood vessel.
(b) The angle '2' is the angle of anteversion or femoral torsion.
(c) The iliopsoas attaches to the bone on the bony process '3'.
(d) The tubercle '4' gives attachment to adductor magnus.
(e) Tensor fasciae latae is one of the muscles which is attached to the bony process '5'.

150 The rectus femoris
(a) lies deep to vastus intermedius.
(b) arises from the femur.
(c) assists in extension of the leg at the knee joint.
(d) is supplied by the obturator nerve.
(e) is bipennate.

149 (a) **True** A branch of the obturator artery enters the head of the femur at this point: other nutrient vessels enter the bone of the neck.

(b) **False** The angle is the 'angle of inclination'.
The neck of the femur does not lie in the same plane as the transverse axis of the lower end of the bone, it is carried forwards.
This angle is known as the angle of 'anteversion or femoral torsion'.

(c) **True** The psoas is attached to the summit of the trochanter and iliacus below and extending onto the shaft.

(d) **True** This is the adductor tubercle: it lies above the medial epicondyle.

(e) **False** Tensor fasciae latae inserts into the iliotibial tract.
Muscles which insert into the greater trochanter include gluteus medius, gluteus minimus, piriformis and obturator internus. Obturator externus inserts into the trochanteric fossa.

150 (a) **False** The belly lies on the superficial surface of vastus intermedius.

(b) **False** The muscle arises by two heads from the pelvic bone, the straight head from the anterior inferior iliac spine and the reflected head from the bone above the acetabulum.

(c) **True** The quadriceps is an extensor of the leg at the knee, but rectus femoris can also flex at the hip.

(d) **False** The femoral nerve supplies all parts of the quadriceps.

(e) **True** Its superficial fibres are arranged in a bipennate manner.

References

Gray's Anatomy (1980). Churchill Livingstone: 36th Ed.

Pegington, J. (1987 and 1988). *Clinical Anatomy in Action* (3 volumes), Churchill Livingstone: Edinburgh.

Cunningham's Textbook of Anatomy, Edited Romanes, G.J. (1981). Oxford University Press: 12th Ed.

Last R.J. (1972), *Anatomy Regional and Applied*. Churchill Livingstone: Edinburgh: 5th Ed.

Hutchinson, M.C.E. (1987), *A Study of the Atrial Arteries in Man*. J. Anat: **125**; 39–54.

Dodd, H. and Cockett, F.B. (1956). *The Pathology and Surgery of the Veins of the Lower Limb*. Livingstone: Edinburgh.

First published in Great Britain 1989

British Library Cataloguing in Publication Data
Pegington, John
Multiple choice questions in anatomy.
1. Man. Anatomy
I. Title
611

ISBN 0–340–50785–3

Typeset in 8½/9½pt Helios by Colset Private Limited, Singapore. Printed and bound in Great Britain for Edward Arnold, the educational, academic and medical division of Hodder and Stoughton Limited, 41 Bedford Square, London WC1B 3DQ by Richard Clay Ltd, Bungay, Suffolk